CONTACT FLYING
Jim Dulin

CONTACT FLYING

Copyright © 2005 by Jim Dulin
All rights reserved

No part of this book shall be reproduced, stored, in any retrieval system, or transmitted by any means, electronic, mechanical, xerographic, audio/visual record, or otherwise, without the written permission from the author. While every precaution has been taken in the preparation of this book, Jimmy A. Dulin assumes no responsibility for errors or omissions. Neither is liability assumed for damages resulting from the use of the information contained herein. None of the material in this supersedes any documents, procedures, or regulations issued by the Federal Aviation Administration.

Cover and all interior design by :
Nyle Douglas Gordon – ndgordon.com

International Standard Book Number: 978-0-6152-0983-8

First Edition

Printed in the United States of America via Lulu Press (www.lulu.com)

Publisher: Jim Dulin, 504 Docker St., Crane, MO 65633

TABLE OF CONTENTS

Foreword   5

Prologue   7

Introduction   33

Chapter 1	Stalls And Slow Flight	40
Chapter 2	The Takeoff	51
Chapter 3	Dutch Rolls	89
Chapter 4	The Energy Management Turn	99
Chapter 5	The Apparent Rate of Closure Approach	106
Chapter 6	The Go Around	124
Chapter 7	The Downwind Turn	128
Chapter 8	The Low Level Forced Landing	136
Chapter 9	Low Power Mountain Flying	148
Chapter 10	The Cross Country	174
Chapter 11	Agricultural Operations	179
Chapter 12	Pipeline Patrol	185
Chapter 13	The New Frontier	197

CONTACT FLYING

FOREWORD

The Flying Fraternity Is A Small One

I first read Jim Dulin's notes in 2002. We had met by chance when his pipeline patrol plane broke down in Moberly, Missouri. I was called to fix the plane. As it was too late to fly after repairs were made, we went to supper together and discovered we had served in Vietnam at the same time but in different units.

As a seasoned flight instructor I knew what it took to make a pilot and when Jim said most of his students soloed in six hours I wanted to know how. He fished a CD out of his overnight bag and said I should read his notes. That CD contained the heart of the material in this book. The flying fraternity is a small one and the ranks of the instructor community are even smaller so, after reading about Jim's techniques and ideas, I felt strongly that his message should get out. I took it one step further and invited Jim to help me train a bush pilot candidate who was going to Africa to be a missionary pilot. I wanted to see first hand if Jim's way of training really worked. I was not disappointed and that young man is now a private pilot gaining cross-country time before he goes on to the mission field.

Wolfgang Langewiesche wrote his classic "Stick and Rudder" in 1944. It still remains a definitive work on piloting, but so much more has been discovered in the past 68 years. Jim Dulin has over 50 years of experience in airplanes and helicopters. Most of the fixed wing time has been in spray planes. Both these aircraft operate in a close to the ground environment. Jim has been able to translate that experience into a logical, learnable training syllabus that gives the student a rapid grasp of their environment and the ability of the aircraft.

CONTACT FLYING

With the introduction of the Light Sport Aircraft there has been a lot of discussion both pro and con about only needing twenty hours of instruction to attain the LSA license. Instructors have been skeptical, and rightly so, on how this can be accomplished when the FAA's practical standards for a private pilot license requires forty hours minimum and the national average for a private pilot certificate is around seventy-five hours in the real world. I believe "Contact Flying" is the answer for how we can train the LSA pilot safely and make him a safer pilot.

If instructors, pilots and students will read this book and practice the contact flying procedures taught herein, we will all become more proficient and better pilots. Whether you fly a Bonanza or ultralight, whether you are a brand new instructor or a "grizzled" old aviator, you will benefit from the techniques put forth in this book. Jim Dulin is not only a gold mine of low, slow flying, but his stories throughout this book illustrate the hard lessons that many pilots have learned but didn't live to tell about. Some of his experiences will raise your hair but he's humble enough to share his mistakes and his failures. I hope you won't read this book and lay it aside. Study the techniques and then go out and practice them to become a better, safer pilot. Enjoy the stories and learn from one of those rare "Old bold pilots."

-Richard Castle January2008

MEII, AP/IA Vietnam 1968, '69

PROLOGUE

I was shot down on my first combat mission in Vietnam. That day I learned that aggressive flying is necessary in a (contact) close air support role. I also learned to be humble. While flying the most sophisticated and most powerful helicopter gunship in the world, I was shot down by a dedicated NVA (North Vietnamese Army) platoon using old and unsophisticated AK-47 rifles.

I arrived in Cam Ranh Bay, Vietnam 03 Nov 1970, spent the night, and got on a Caribou fixed wing transport plane to Bien Hoa. It seems the First Air Cavalry Division, to which I had been assigned, had picked up and moved from An Khe in the Central Highlands to Bien Hoa in III Corps and my orders had not caught up to the mobility of an air cavalry division. Actually I had received orders for Vietnam with TDY (Temporary Duty) enroute to flight school more than a year previously. Except for a very few Army National Guard officers, everyone went to ORWAC (Officer Rotary Wing Aviation Course) on Vietnam orders. After five days at FTA (First Team Academy) the in-country refresher school, I checked into the famed 1/9th (pronounced First of the Ninth) Air Cavalry at Pouch Vinh during a rocket attack. The 9th Cavalry Regiment had been the famous all black "Buffalo Soldiers" against Geronimo in the Southwest. The Apaches honored them with the name "Buffalo Soldiers" because their hair resembled that of the honored buffalo and they were tough fighters. The 1st Air Cavalry Squadron of the old regiment had again become famous and highly respected by the enemy in Vietnam.

These rocket attacks were simply harassment tactics like the mortar fire into the Green Zone (protected area in central Baghdad, Iraq) today. The VC (Viet Cong or Victor Charlie in the military phonetic alphabet shortened to "Charlie") launched rockets from bamboo tripods with water drip buckets and clothes pin delays so they could be back in town when we bounced a Cobra helicopter gunship to deal with them. After

CONTACT FLYING

investigating a LOH (Light Observation Helicopter pronounced "Loch" as in roach with l replacing the r) crash back at Bien Hoa, I checked into Alpha Troop in Song Be near the Cambodian border late on the 11th. Late on the 12th I was back at Bien Hoa as a patient at 93rd Evacuation Hospital.

I was a "new guy" only one day in Vietnam. I arrived late enough in the day at Song Be that I was not scheduled to fly on the 12th. Of course I talked CW2 (Chief Warrant Officer Second Grade) Monte Johnson out of his copilot seat on an AH1-G (Assault Helicopter Series 1 Model G or Cobra) with "Bloody Bart" as AC (Aircraft Commander). In other units, as a 1st Lieutenant, I would have been AC as I outranked CW2 Bartlett but The Air Cav wasn't being stupid. The 1/9th Air Cavalry put experienced people in charge. I was a "new guy" only one day because we got shot down that day and getting shot down makes you an old guy.

Mr. Bartlett (Warrant Officers are properly addressed as Mister), a railroad man from Montana, got his name from his habit of getting into situations that involved making boasts and throwing bad phrases around. Bart's was "bite my ass." Somehow this phrase evolved into not giving a rat's ass and finally into a boast about biting a rat's ass. I'm not sure as war stories flavor with age, it's been a long time, and I got the story second hand. Anyway, Bart ended up biting a rat's ass.

We had rat boxes everywhere for sanitation and sport. The Blues (the one infantry platoon in each Air Cavalry Troop was called the Blue Platoon or The Blues) removed the grenade projectiles from their M-79 Grenade Launcher ammunition. With the powder and plastic wadding only, they shot rats as they were released from live traps. "Pull"! Boom! Blood and guts all over the wall, is sort of how it was done. Someone once said Vietnam was a live fire training action with a lack of adult supervision.

As stories and situations evolve, Bart had to retell and of course demonstrate the biting incident from time to time. The pilots naturally began to refer to him as "Bloody Bart." Oh! I almost forgot. There were the songs with Bart's poetry and Johnny Cash tunes: "Six hours in the air and I'm going to make it home tonight," "Napalm sticks to kids," stuff like that. We weren't crazy, just alternately bored and scarred to death. It's called combat.

On the 12th I got up early and went down to the flight line to check my assigned Cobra gunship. First Cav aircraft were notoriously rough but I didn't know what good Army maintenance was at that time. I had only seen training aircraft, which are rough, and this one. I did wonder aloud about the lack of mini-gun and chunker (electrically operated 300 rounds per minute grenade launcher in the turret) ammo and was told neither weapon was operational and that the armament people were flying door gunner rendering them unavailable. I was going as copilot but not as copilot/gunner this day.

On the way out to the AO (Area of Operations) Bart explained troop TO&E (Table of Organization and Equipment), lessons learned in Vietnam, and Alpha Troop SOP (Standard Operating Procedures). Air cavalry troops were completely organized around the helicopter. There were ten AH1-G Cobra gunships in the Gun Platoon (Red Platoon). The Scout Platoon (White Platoon) had ten OH6-A LOHs. The troop also had one infantry platoon (The Blue Platoon) carried into battle by ten Blue Platoon UHI-H troop hauling helicopters. This utility helicopter called a Huey carried no fixed armament. We gunnies called them slicks which meant without guns. We kidded that, "slicks are for kids" knowing that they were braver than us because they were willing to enter a hot LZ with only door gunners.

Standard Operating Procedures were simple and effective. We generally operated as a Pink (Hunter-Killer) Team made up of one Cobra gunship, called a Cobra, gun, or high bird from the Red Platoon and one scout ship, called a scout, Loch, or low bird from the White Platoon. Red and White make Pink so a Cobra gunship covering a scout Loch was called a Pink Team. *See Figure 1*. When the Blues were on the ground we were a

Purple Team made up of a Red (Gun), a White (Scout), and the Blues (The Blue Platoon).

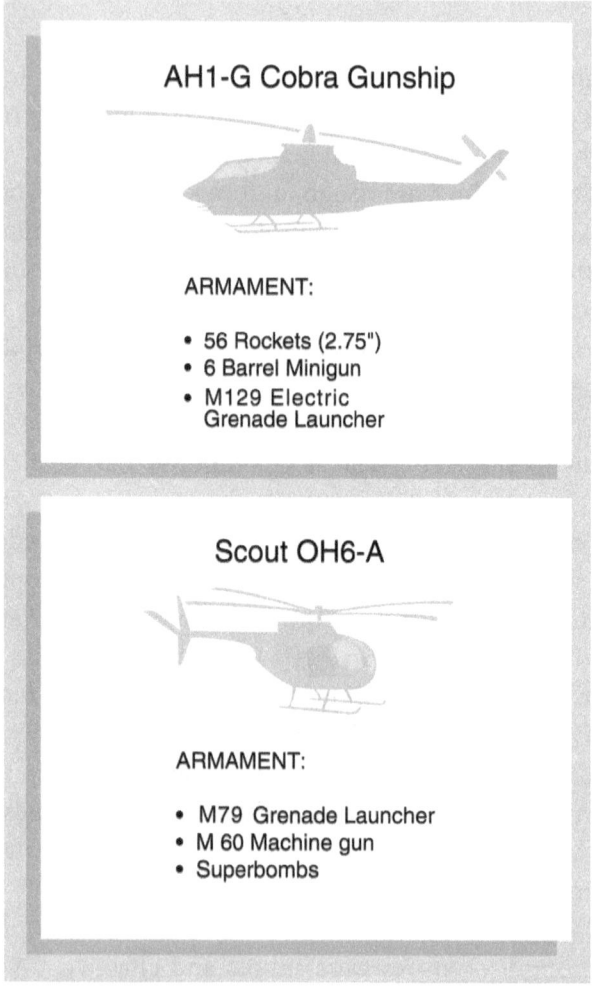

Figure 1

The scout would spiral down from 3,000' to the deck and begin circling to the right (the pilot sits on the right in a helicopter) at about 40 knots in

the songs with Bart's poetry and Johnny Cash tunes: "Six hours in the air and I'm going to make it home tonight," "Napalm sticks to kids," stuff like that. We weren't crazy, just alternately bored and scarred to death. It's called combat.

On the 12th I got up early and went down to the flight line to check my assigned Cobra gunship. First Cav aircraft were notoriously rough but I didn't know what good Army maintenance was at that time. I had only seen training aircraft, which are rough, and this one. I did wonder aloud about the lack of mini-gun and chunker (electrically operated 300 rounds per minute grenade launcher in the turret) ammo and was told neither weapon was operational and that the armament people were flying door gunner rendering them unavailable. I was going as copilot but not as copilot/gunner this day.

On the way out to the AO (Area of Operations) Bart explained troop TO&E (Table of Organization and Equipment), lessons learned in Vietnam, and Alpha Troop SOP (Standard Operating Procedures). Air cavalry troops were completely organized around the helicopter. There were ten AH1-G Cobra gunships in the Gun Platoon (Red Platoon). The Scout Platoon (White Platoon) had ten OH6-A LOHs. The troop also had one infantry platoon (The Blue Platoon) carried into battle by ten Blue Platoon UHI-H troop hauling helicopters. This utility helicopter called a Huey carried no fixed armament. We gunnies called them slicks which meant without guns. We kidded that, "slicks are for kids" knowing that they were braver than us because they were willing to enter a hot LZ with only door gunners.

Standard Operating Procedures were simple and effective. We generally operated as a Pink (Hunter-Killer) Team made up of one Cobra gunship, called a Cobra, gun, or high bird from the Red Platoon and one scout ship, called a scout, Loch, or low bird from the White Platoon. Red and White make Pink so a Cobra gunship covering a scout Loch was called a Pink Team. *See Figure 1.* When the Blues were on the ground we were a

CONTACT FLYING

Purple Team made up of a Red (Gun), a White (Scout), and the Blues (The Blue Platoon).

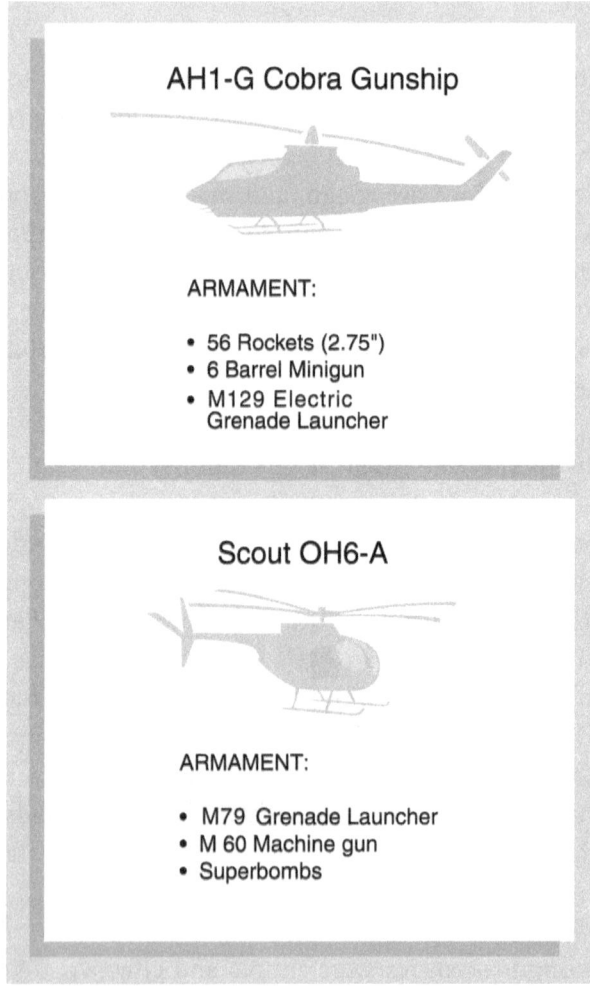

Figure 1

The scout would spiral down from 3,000' to the deck and begin circling to the right (the pilot sits on the right in a helicopter) at about 40 knots in ground effect in LZs (all open areas were called Landing Zones). Over the trees he moved slower to see into the canopy or blow bamboo away with his rotor wash.

The Red Platoon Cobra covered the low bird (see illustration above) in a lazy circle at twelve hundred feet AGL (Above Ground Level), about forty knots airspeed, and twenty two pounds of torque (the engine torque at which it was easiest to trim a Cobra to insure that rockets are accurate). When pairs of rockets leave the wing stores, they align with the relative wind. If you're out of trim they will be inaccurate. The Cobra AC or back-seater kept the low bird in sight at all times or would call him up immediately if he lost sight of the Loch.

The Gunship AC flew the Cobra, kept the low bird in sight while communicating with him on UHF (Ultra High Frequency radio), shot rockets, and ran the mission unless the "Old Man" (Troop Commander) was around in his Loch or a C&C (Command and Control) Huey. The front-seater (copilot) was a very busy man. He covered the break from a gun run with the turret (mini-gun and chunker). He navigated, which in III Corps meant keeping his finger on the map at all times no matter what else happened. He communicated with Div Arty (1st Air Cavalry Division Artillery was airmobile and kept our ground troops continuously covered) and the Blues on FM (Frequency Modulated radio). He communicated with the FAC (USAF Forward Air Controller) on VHF (Very High Frequency radio). He copied the spots (spot reports called up by the scout pilot) onto the gunship canopy with grease pencil and called those spots back into the troop TOC (Tactical Operations Center) on the way home from a mission. The Commanding General of the First Air Cavalry Division once said half of the fights the Division got into were initiated by the First of the Ninth. They wanted those spots immediately.

Our scout pilots, like the Indian scouts of the horse cavalry, could judge the number of enemy that had used a trail, how recent the use, find spider holes and breather holes (for underground bunkers), campfires, clothing,

CONTACT FLYING

etc. The observer in the left seat of the Loch held a red smoke grenade with the pin removed in his hand at all times while working. The torque (M-60 machine gunner) sitting on the rear compartment floor with his feet on the right skid reconned by fire (shot into suspected enemy hiding places) or dropped super-bombs (M-60 ammo can filled with C4 plastic explosive and primed with a concussion grenade) into bunkers. He also carried the M-79 grenade launcher and assorted hand grenades.

Upon receiving enemy fire, the scout pilot either slowed up to "check it out" or bugged out. *See Figure 2.*

The scout pilot would call "taking fire, coming up" (or "clearing west" etc.) and the observer would drop the red smoke grenade. Flying or crewing a Loch in a scout platoon was the most dangerous job in Vietnam. A much overused but accurate statement in Vietnam was, "the LZ is marked by the burning Loch." Causalities were very light compared to WWII or Korea but the two troopers killed in my troop (about sixty pilots and one hundred men) during my tour of duty were a scout torque and a scout pilot in separate engagements.

As the Cobra was already nose up at 40 knots when the scout called, "taking fire," the gunship back-seater simply rolled onto the target (red smoke) banking usually 120 degrees (the same energy management turn as the old lazy eight or crop duster turn) allowing the nose to fall through naturally to a sixty degree pitch down dive. Even if the scout stayed on target and we did not fire, the double whop-whop sound of the blades of the 540 rotor system (which made the Cobra an effective gun platform) in a tight turn caused the bad guys (enemy) to get their heads down and quit firing at the scout. When the scout cleared the area, the Cobra AC fired pairs of rockets onto the red smoke from the smoke grenade.

Figure 2

CONTACT FLYING

The scout usually went back in to check things out but the Cobra frontseater (copilot) could call artillery and the FAC usually rang in wanting to put Zoomies (fast moving fighter bomber jets) in. Finally if the quality of the find warranted it, the Blues were lifted right onto the spot where the observer threw the smoke.

The Blues didn't walk around in the woods, they rode Hueys and fought. In hot areas we might operate as a Purple Team (Gun, Scout, and Hueys with Blues) from the beginning. For the enemy, shooting at an air cav helicopter would bring a lot of helicopters and a lot of trouble. When anyone in The First Air Cavalry was hit, all you heard was hitting back. We hit them back with more than ten times the firepower.

Every unit in Vietnam, whatever their tactics, respected the "Lessons Learned in Vietnam" that were a part of our indoctrination. In Apache Troop the lessons learned were the combined experiences of all the troopers who had been in the First of the Ninth during the five years the squadron had been in Vietnam. We were to never fly single ship or to follow another helicopter in trail formation, or to fly over the same place twice. We were not to over fly the target nor were we to fly between 3,000' AGL and 100' AGL (dead man zone). Low level (50' AGL) was safer against large caliber weapons (mainly 50 cal.). We used everything that would fly (even maintenance grounded aircraft) when a helicopter was down or the Blues were in contact with the enemy. We always keep a gunship over the Blues. We got everybody out dead or alive (most shot down crews were picked up by another helicopter immediately). If captured we were to escape and evade at the first opportunity. We were to cease firing when the good guys (us) were down and their location was not fixed.

On the 12th Bart was giving me an area checkout and covering the Blues while their assigned Pink Team cover rearmed and refueled. We had ample fuel for a short diversion when relieved and this lured us enough that we failed to abide by two of the lessons learned: Do not fly a mission single ship and do not over fly the target. The most exciting day in my life started with Bart's bold cavalry declaration, "I'll show you where some bad guys are."

After the First Air Cavalry Division had successfully spearheaded the short strike into Cambodia that produced the largest weapons and ammo caches of the war, the NVA (North Vietnamese Army) had ordered its units in the south to be responsible for their own logistics and return to a less combative level of insurgency. They were forcing Montagnards to farm small plots for them in the Song Be Mountains. After entering this enemy rich area low level and single ship, we caught a NVA Platoon (khaki uniforms with pith helmets) in the middle of a large LZ.

Bart pulled up, with enough force that my head hit the armored headrest, and rolled in on them. Next I was looking at an incredible scene that seldom occurred in Vietnam: NVA regulars caught in the open. *See Figure 3*. They were not running as were the VC and the Montagnards. They were staying put in the middle of that large LZ. I had read about the tactic unfolding before me but I hadn't believed any unit was gutsy enough to pull it off. This is the drill. The platoon leader orders his troops to take a knee and place their AK-47s on full automatic pointed skyward. He waits until the gunship is almost on him in its gun run. On his signal they fire. This is a very bold tactic when undertaken with rockets impacting into people all around. On our third gun run we had gotten too low and too slow and yes we over flew the target.

CONTACT FLYING

Figure 3

The police say that if they interview three witnesses to an incident they will get three separate descriptions of what happened. In June of 2005 I located Bart and visited him in Montana. I found out that Bart had been chasing one of the VC who ran up a streambed. The NVA, stationary in the middle of the LZ, he had not seen. I had seen them but couldn't shoot. I had never thought, during the few days we were both in Song Be after I returned from 93rd Evac, to ask him if he had seen them. Combat, when it gets up close and personal, is a very confusing endeavor.

I saw the needles split (see *figure 4*) and heard the rotor RPM low audio alarm as Bart bottomed the collective (took all pitch out of all blades with the collective) to maintain rotor rpm.

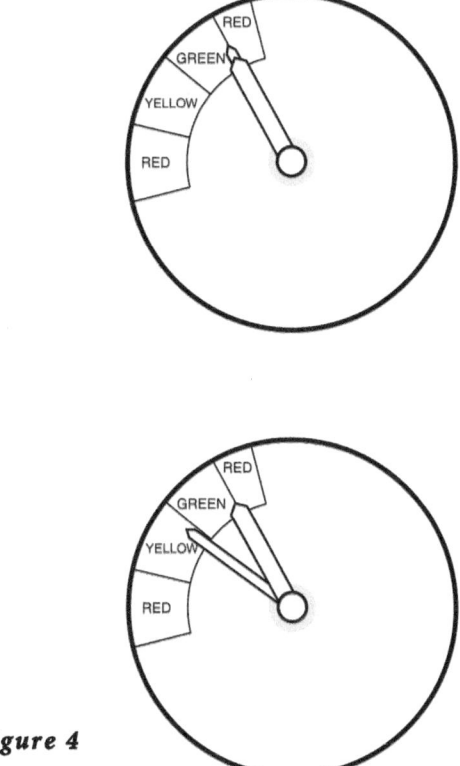

Figure 4

CONTACT FLYING

During normal operations turboprop jet engines operate at 100% n1 (gas producer) which is full power development. That produced hot gas flows over the n2 power turbine which is attached to a propeller in airplanes or a rotor in helicopters. In airplanes the pilot uses that already produced power by feathering or changing the pitch on his propeller with a lever that acts much like the throttle in a piston airplane. In helicopters we do the same thing with the collective pitch lever. The collective changes the pitch in all blades equally whereas the cyclic changes the pitch in only one blade at a time through the swashplate. The swashplate changes any non-rotating control input (collective or cyclic) into rotating control input. On the RPM gauge the thinner and longer n1 gas turbine speed needle is superimposed over the fatter and shorter n2 rotor speed needle. When n1 is the desired 6600 rpm and n2 is the desired 340 rpm both needles are joined. This arrangement allows the pilot to quickly determine that all is well with the world. When the engine fails the narrow and long engine RPM needle falls off. At the same time the rotor slows causing the rotor RPM needle to fall off into the yellow range setting off the rotor RPM low audio alarm, the master caution light, and the rotor RPM low warning panel light. This is because we still have pitch in the blades but no n1 gas to drive the n2 power turbine which turns the rotor. The audio gets your attention and you must reduce the pitch in all blades with the collective or you will lose turns and die. When you slam the collective all the way down, inertia speeds the rotor back up and you may have to pull some collective pitch especially in a turn. All this happens in less than a second and when you look at the RPM gauge the needles are split (engine n1 needle is counterclockwise of or below rotor n2 needle). Bart auto-rotated through the sparse 200' tall trees we called Loch eaters and for 36 years I thought he had flared on top of the main jungle canopy (triple canopy jungle).

The last thing I remember is wondering whether Bart was going over or under a large limb sticking out horizontally from a Loch eater. For 36 years I had believed that once slowed to a stop on top of the main jungle canopy, we had fallen one hundred feet to the jungle floor. I didn't get a chance to go discuss the crash in detail with Bart until 2005 when I found him back at his homeland at Whitefish, Montana. He had found a hole in the second canopy and put us down on the bamboo going sideways to slow down. My Mickey Mouse helmeted head had impacted the canopy.

The height limit for a Cobra pilot was six feet. Even at five feet seven inches my helmet had sometimes vibrated against the canopy in perfect harmony with the whap whap of the rotor. It was a tight fit up there.

Bart, thinking I was dead, dragged me into the brush so they wouldn't mutilate me. After seeing the Loch eater limb, the next thing I remember seeing was the Cobra on its side with rockets hanging surreally from the rocket pods as if they had dripped partially out. It was what I heard, however, that scared me. I heard the sound of whispers in Vietnamese and the quiet movement of disciplined soldiers.

I was dead. I had left Song Be that morning flying the world's fastest and best armed helicopter gunship. I was a member of one of the most famous aviation units (the same air cavalry unit you saw in <u>Apocalypse Now</u>) in the most powerful army on earth and I was dead. I had a 45 Cal. automatic pistol and two nine round clips. I had no survival radio like aviation units further from the action carried on their survival vest. Nor did I have the survival vest. I had been knocked out and I had lost a lot of blood from smashing my leg into the gunner's quadrant. I knew enemy medical attention and food sufficient for recovery would be time consuming and wasteful of meager supplies. I was facing the shaken but dedicated survivors of the thirty man platoon Bart had decimated. They had AK-47s and they had recently demonstrated the courage to use them. I was dead.

I prayed the Lord's Prayer and promised a major change in outlook if spared death. I buried my SOI (Signal Operating Instructions) and I ran. Fortunately the Song Be Mountains were very steep and the NVA were uphill from me as I fell with every step but made good progress anyway. I found myself at the edge of a large LZ just as a scout flew over. I ran into the LZ and waved my blood chit (American flag with promises to pay for my safe return in several languages) which was the only thing I had not colored OD (Olive Drab). Scout pilots in the troop later accused me of trying to chu hoi to the NVA. VC often waved leaflets dropped by Army

CONTACT FLYING

Psychological Operations that offered them money and amnesty to change sides or chu hoi in Vietnamese. Then a Huey came to a hover over me. Mr. Jackson, a shipless slick pilot who had just jumped in a Huey when the downed aircraft alarm went off at Alpha Troop, had to jump out and boost me up because the slope was too great to allow a normal hover. I could barely reach the skid with my outstretched hand. The door gunner jumped out and retrieved the pistol I had just tossed onto the ground. These were sharp troops.

When we were overdue, based on fuel endurance, and assumed down the Blues had been picked up to operate as a recovery Purple Team. I had not gotten a call off to Apache Troop but they knew how Bart operated. While I was being Med-evaced they were rappelling onto the broken Cobra under enemy fire. You gotta love those Blues. Bart had headed out in the direction he mistakenly thought was the LZ that I had ended up in and had run right into the NVA Platoon coming up the mountain. He had hidden out as they passed him in a deployed assault line formation. When he heard the slicks with the Blues he crawled back up the hill through the retreating NVA Platoon. He yelled, "don't shoot Blue Mike (the Blue Platoon Sergeant's call sign) its Bart," and ran into their hastily established perimeter.

I returned to Alpha Troop after a week at 93rd Evacuation Hospital. They would have sent me on to Japan and then reassigned me but I had gotten a doctor to sign me out one night. I was on pain killers for another week and couldn't fly; so I got my hooch mate, the Scout Platoon Leader, to swing me a Loch checkout and I flew as observer with the scouts. When back on official flying status, I went back to guns; but I never took a round in the aircraft nor did I lose a scout the rest of my tour.

After the crash, Bloody Bart was kept off flying status the seventeen remaining days of his tour. He drank heavily most of those days and through his stories I learned who he was. As Colonel Hal Moore said of his First of the Seventh Cavalry troopers in the famous Ira Drang battle, "we were soldiers once, and young." John Bartlett, alias Bloody Bart, was typical of our very good small unit leaders in Vietnam where platoon sized engagements were the norm. He led by example in the way of the Norsemen of old. People liked doing what Bloody Bart asked them to do. When he looked over his shoulder in battle they were always back there. John Bartlett was a great pilot and a superb leader. We were, all of us, good soldiers; but we were never again young.

What we at Contact Flying and Aircraft do and teach is dangerous enough that the grace of God is required to survive the learning curve. But this is true in all aspects of life. Many see pride in Alfred Lord Tennyson's "ours not the reason why, ours but to do and die." I see great faith in <u>The Charge of the Light Brigade.</u> The Marine motto, "Do or die." has great pride. But in doing and dying Tennyson's Light Brigade accepted the fact that God is truly in control and thus our effectiveness or "blunders" have less bearing on the situation than we think. That two hundred of the six hundred troopers survived the charge was providential.

Pilots are proud people, and I am an old bold one. But pride is expensive. According to Solomon, pride goes before the fall. Pride must be tempered by humility for true boldness to mature. My greatest concern is that my boldness and teaching of it reflects my faith in the creator's love for all men. True boldness is a gift from the Almighty, not a product of man.

In education training at University of Missouri at Kansas City, I learned from Professor Gwynthomas that it is unwise to attempt to teach a lesson or unit we cannot justify. We must explain why what we teach should be taught. If all flight training requires immediate instrument integration and the preparation of all beginning students in ways that promote the transition to larger aircraft, how are we to justify primary and advanced instruction in small, low powered, simple aircraft? Why should my company, Contact Flying and Aircraft, even bother with, much less emphasize, contact flying? For those of you who are airline or corporate aircraft bound, we cannot promote or defend a program of instruction

CONTACT FLYING

that appears to be illegal, dangerous, and ludicrous. Where the management of large, jet powered, computer guided aircraft is involved, the art of flying by contact with the ground using all of our senses without electronics is not a required or appreciated skill. But if you are willing to work long hours in bad weather using your eyes to see small mile markers on pipeline patrol, or to feel the energy bleed of a crop duster aircraft coming out of a field; or to decelerate below stall speed (behind the power curve) in order to haul hunters into very short strips you can learn this art called flying. If you can accept working so close to the ground that concentrating on any instrument is suicidal, you can learn contact maneuvering. If you can learn to work with, rather than against, your aircraft using the power of the Almighty through his natural laws, which far surpasses the power of man through his engine, you can become a better pilot. If you can fly fluidly and dynamically rather than rigidly and statically you can fly with greater confidence. If you can use all your senses rather than just the sight of certain numbers on gauges, you will be more competent. If you can accept a brisk walk apparent rate of closure with the ground over a "stabilized approach" on instruments, you can fly at a level of skill many pilots never achieve. The Contact Flying and Aircraft organization can't make you an old, bold pilot but we can give you a contact flying experience that will make you competent enough to approach difficult flying situations with a boldness based on that experience.

Many are drawn to flying because of the rigid, regulated, right stuff mentality portrayed by professional flight schools, just as many are put off by this attitude. Many good pilots are driven from the industry because they fly flexibly rather than rigidly. This culture of regulation is present in both the peacetime military and civilian aviation. Air cavalry pilots, like crop dusters are not always considered professional. Ag pilots can teach crop dusting and the Army can teach scouting, troop hauling, NOE (Nape of the Earth) flying, medical evacuation, and providing gun cover. Contact Flying and Aircraft offers instruction in true contact flying which is sorely lacking in an instrument oriented program. I have great faith that the new Practical Test Standards for the Light Sport Aircraft Pilot Test will help our school in this endeavor. These standards have more flexibility than the old Private Pilot and Commercial Pilot Practical Test Standards.

At Contact Flying and Aircraft we call flying solely by reference to the ground contact flying. Our vision is directed totally outside the aircraft. Our senses of smell, feel, and hearing augment our visual cues but our scan is 99.9% outside the aircraft. Once sanctioned and taught by the establishment, contact flying has been allowed to lapse in today's approved instructional program and the Private Pilot and Commercial Pilot Practical Test Standards. Our poor general aviation safety record does not reflect on students and instructors but on the establishment that insists that small, low powered aircraft fly like large, jet powered aircraft.

Why is the accident rate higher for low powered aircraft? Could it be that our standardized training program (oriented toward large, computer guided, IFR aircraft) poorly teaches the art of flying small aircraft by reference to the ground? Why can't we choose to initially train those aspiring to corporate and/or airline flying in large aircraft at large airports? Why can't we train all other students in contact flying in small trainer aircraft at small airports? Why isn't <u>Stick and Rudder</u> the standard for all primary instruction? Will sanctions against those who err clear up the confusion resulting from training in little airplanes as if they were big airplanes? Am I just a crazy old crop duster or are these proper considerations for curriculum development in the 21st Century?

There are those in flying who value field experience, both combat and civilian. After crashing my Pawnee Ag plane midway through my second season, I had to go hat in hand to lease one at a reasonable rate. I told the old crop duster I needed one of his Ag airplanes because I had just crashed mine. Asking neither for license nor logbook, he was quite comfortable with leasing me a Callair Ag plane until he learned I had only crashed once. At this point he wasn't sure I had enough experience to lease his

CONTACT FLYING

plane. In Vietnam I never had to call for air support because it was offered air cavalry hunter killer teams without solicitation. When my low bird started finding enemy clothing, recent campfires, etc. the Air Force FAC (Forward Air Controller) offered "Apache 21 you're going to find some bad guys. I've got a stick (flight of four) of F-4s coming out of Tan Son Nhut in zero five. I'll put them in an orbit two miles west. Let me know when you need them."

Most segments of our industry value only those who have never erred. But, like old crop dusters and Vietnam FACs, there are those who value the down to earth (literally) experience not reflected in logbooks. There are many schools that teach the science of flying by the numbers with a never fail dogma and we need them for the resource management pilots. We can not teach future airline pilots that Vx does not exist in the mountains because it steadily accelerates to Vy as you ascend. The idea that anything but a violent pitch up will result in smacking a sky scraper must be brainwashed into them. Contact Flying and Aircraft will put you with old bold pilots who have survived and can expand your envelope with their hard earned knowledge. Our school provides a flight training program for those who live, work and play outside the corporate/airline box.

The loss of the 737 at National Airport must be considered an acceptable loss (he tried to go up when the airplane would not) but not so the Cardinal at Cheyenne Wyoming, where a budding young pilot lost her life. The seven year old girl and her instructor from California, on a highly publicized transcontinental flight, need not have believed the elevator always elevates and the engine never fails to produce needed power. Vastly better it would have been had they understood two essential and instinctive questions asked by mountain pilots in every tight situation. We always ask ourselves which way is downhill? Next we evaluate any obstruction that would prevent flight in ground effect and/or a descent to get there. What I have learned about the art of flying has come through experience. My school teaches the art of flying from the collected experience of bold pilots who have survived dicey situations.

I have had many forced landings, all resulting from one or more errors on my part. Can I ethically and legally share this inappropriate knowledge? Can the general aviation community benefit from students, instructors, and old bold pilots who do not believe, because of the uncomfortable things they have experienced, that small airplanes fly like very powerful airplanes?

Other than to provide Contact Flying and Aircraft students with a text, my objective is to encourage <u>Contact Flying</u> readers to "test the spirits" as the apostle John recommended. And here I am with a commercial pilot license for airplanes and helicopters, instrument certificate for airplanes and helicopters, instructor certificate for contact and instrument operations, agricultural operating certificate, and a logbook showing over 16,000 hours of flight time. Yet no piece of paper proves that I know anything about flying. Paper does not make pilots safer or more proficient. Experience with actual precautionary landings, forced landings, combat operations, medical evacuation operations, bush operations, agricultural operations (crop dusting), and mountain flying operations have made me a safer and more proficient pilot. Flight Safety and other schools have excellent programs for corporate and airline pilots but we are one of very few civilian contact flying schools. We all have our own point of view and we are all full of it. This is the nature of man. At Contact Flying and Aircraft we attempt to present the contact flying point of view.

We are taught the science of flying by a bureaucracy that tolerantly invites public comment when codifying rules and training standards but once those rules and standards are finalized, the FAA intolerantly sanctions nonstandard techniques. This makes our Instrument Flight Rules the worlds safest and in fact the world has pretty much adopted them. My concern is for contact flight operations under Visual Flight Rules. The

CONTACT FLYING

Practical Test Standards for the LSA Pilot, Private Pilot and Commercial Pilot certificates do not properly address all contact flight operations. Especially in the "maneuvering flight" area, our general aviation safety record is poor and I believe it can be improved with the acceptance of nonstandard operating practices and instructional flexibility where appropriate. This may take time and there are important changes in the LSA standards and but much empirical evidence is available to any who accept that small, low powered aircraft fly differently than large jet powered aircraft.

This book and our school attempts to teach the art of flying small, low powered aircraft by reference to the ground. The expressed theories and actual practices attempt to follow Federal Air Regulations and Practical Test Standards. Where these regulations and sanctioned training techniques and parameters conflict with reality and hard earned experience, however, practical theories and techniques based on contact flying are taught. The tolerances in the LSA PTS, that give common sense leeway from rigid V speeds, show promise in this area.

Wolfgang Langewiesche and I are the only pilots I know whom openly differ with politically correct dogma. Every pilot should read <u>Stick and Rudder</u> by Wolfgang Langewiesche. He explains the theory of flight from the pilot's seat, not form the design engineer's drafting board. Reading his book affords us a better grasp of our art than all of the aviation texts and articles written since its publication in 1947. <u>Contact Flying</u> and Contact Flying and Aircraft try to make practical application of Mr. Langewiesche's very useful theory.

While "trial by oath" originated with "The Church" in the Dark Ages, its present use can get in the way of Christian attitudes and responses in aviation today. I had just landed at an airport and was paying for fuel when a low flying complaint was called in to the FBO. The young lady working that afternoon said she would talk to the manager about it. Because I had just followed a yellow 172 to the pump I explained that the complaint was probably about my yellow 172. I had just come from a pipeline patrol. She was concerned that it may have been their instructor. I assured her it was I based on the area of the complaint but she would not discuss the situation with me. The next morning I observed the instructor and a student getting the yellow 172 ready for a lesson. As a different lady was now in the office I again explained the possible mix up. Neither the office lady nor the instructor would discuss it with me. My conclusion is that tensions resulting from positions and licenses (high and low) prevented anyone from responding to my simple conciliatory statement. This attitude is unfortunate but understandable in a highly sanction oriented culture. If the complaint was from an Airline Transport Pilot, under "trial by oath," my confession was worthless. In the enforcement culture of the FAA, a Commercial Pilot is always trumped by an ATP.

There are many stories out there, many opinions, and many beliefs but there is only one FAA sanctioned method for flying all aircraft. That there are few differing opinions on how to fly airplanes, I believe, has more to do with fear of sanctions than with lack of opinions. <u>Contact Flying</u> and Contact Flying and Aircraft offer non-sanctioned theories, opinions, and techniques. Our methods will not appease "politically correct" pilots. This little book contains over forty years of sweat and blood experience including a tour in Cobras in Vietnam where I left some blood. It should be of interest to those bold enough to even consider spraying crops, patrolling pipelines, flying bush, flying light sport airplanes, mountain flying, or picking injured people up from unimproved remote sites, all legitimate pilot careers. Our state and federal agencies should put our interests first at all levels. We pay the wages of their employees. Present day "trial by oath" is archaic and promotes neither progress nor safety. The guy with the most paper in the form of licenses, certificates, ratings, and titles of nobility is not necessarily right on every issue. I am pleased, however, with the LSA Practical Test Standards and the FAA's willingness to listen to people who understand small aircraft. This is positive and hopeful. Enjoy the book. Test the spirits. God bless you.

The author standing by a revetment at Tay Ninh. Several A Troop 1/9 Cav pilots were injured by the breaking of a rapelling rope used in a tug-of-war. This rope will stretch 120% prior to breaking then return rapidly to normal length. Those of us between the break and the mud hole were injured when the rope ran rapidly through our hands.

Cobra pilot, Lt. Cromar (on wing store) and Browns at Tay Ninh.

Cpt. (Al) Capone in front seat of Cobra at Tay Ninh.

CONTACT FLYING

Hootch and bunker fires started with white phosphorus rockets from an Apache Troop Cobra in Me Kong Valley.

I lost transmission fluid in Cambodia. My low bird had extra on board the Loch. In `71 we were allowed to land in Cambodia only in an emergency.

Slick drivers, (David Brown ?) and Larry Jackson in an Apache Troop Huey.

South Vietnamese girls in Siagon selling rice bread.

In front of my hootch in Phu Loi with 3/17 Cav.

INTRODUCTION

I have not always been a contact flyer. The first seven years of my flying career involved instrument flying only. No integration of instruments with contact flying, I simply flew enroute totally by reference to instruments. The reason for this was that, as a very short nine year old, I could not see over the instrument panel of a Cessna 180. A contributing factor was that my dad's business partner, Mr. J. Press Maxwell, the golf architect, always made me file and fly IFR. He had flown B24s out of North Africa and Italy during WWII. Press flew the C-140, C-170, C-180, Piper Comanche, Piper Twin Comanche, Piper Aztec, and finally the Piper Seneca II to and from negotiations for golf course work. I accompanied him as pint sized copilot/valet. I was the copilot who "squirted" the dog. Mrs. Maxwell had show dogs we often transported.

I received the FAA Commercial Pilot Certificate during college with Army help. I was the first ROTC (Reserve Officers Training Corps) cadet to receive a Commercial Certificate in the ROTC Flight Program. I had requested an enabling waver from DA (Department of the Army) at the beginning of my senior year. The approval came early in June. I flew the thirty-six hours and got my certificate before graduating and being commissioned 24 Jun 1969. The commercial program was not so complicated back then and it did not include instrument training.

In Army Aviation during the Vietnam War, we learned to use the NDB (Non Directional Beacon) in the tactical instrument course at Ft. Rucker, Alabama. NDB transmitters were set up every ten miles to simulate terminals on a cross country flight. Since neither Army helicopters nor most combat theaters had VOR (Variable Omni Range) navigation capability, NDB was our primary navigational aid. We did have a very fine RMI (Radio Magnetic Indicator or slaved compass), however. Like most old Army pilots, I prefer NDB and might even choose to shoot a VOR approach on the #2 needle on the RMI rather than on the CDI (Course Deviation Indicator). I was never comfortable with the loss of positional awareness on the CDI or HSI (Horizontal Situation Indicator) during procedure turns and holding in civilian aircraft (no RMI). Unfortunately many GPS (Global Positioning Systems) also simulate the HSI rather than the simple direct bearing presentation of the NDB or RMI.

After a year in Vietnam and a year in Germany, I joined the 635th Assault Helicopter Company of the Missouri ARNG (Army Reserve National Guard). We were flying out of the civilian field at Warrensburg, Missouri but soon moved to Whiteman AFB (Air Force Base). The Strategic Air Command had given all it's B-52s to the Air National Guard and Whiteman AFB was about to lose flight operations personnel at the base. They gave the 635th three B-52 hangers and all the spare parts that were compatible with our Mike Model gunships (UH1-M or Utility Helicopter series 1 model M).

Army National Guard units set up two week Standard Instrument Courses for Army Aviators who had only Tactical Instrument (NDB) Authorization. Two weeks FTTD (Federal Temporary Training Duty) was allotted to learn VOR navigation and VOR approaches. My stick buddy (students assigned to one instructor are called a stick) was our Company Commander, Major Woods, a dentist from Nevada, Missouri. Our instructor, CW2 Joe Stokes, was a weekend warrior who lived out of his old Volvo station wagon because the Guard was his only livelihood.

Joe Stokes, now married and with the FAA, was our confirmed bachelor and professional comedian. He always had a throng of listeners hanging on to his every word as he told well oiled war stories. He was a master of timing and knew how to work an audience. He had been a maintenance officer in a Chinook (twin rotor heavy lift helicopter) outfit in Vietnam. He didn't have much good to say about Chinooks. His favorite yarn was about a Chinook pilot who tried to fly under a high voltage electric transmission wire somewhere in Texas. He would get this story started and then go on to another equally interesting story. During the second story someone would always ask (he knew they would) what happened to the Chinook pilot? Without so much as a pause Joe would say, "He crashed and burned," and continue with the second story.

For our standard instrument training we loaded up in a Mike Model (UH1-M) gunship and, as Joe explained it every time when various airport workers asked, "We're just following the clouds." We followed the clouds to Chicago O'Hare International Airport and made many VOR and ILS approaches to minimums with a student (the Major or I) and Joe up front. On the way back in good weather, however, Joe used the jump seat between the Major and me. This is the usual Army practice for the instructor in VMC (Visual Meteorological Conditions).

An interesting thing happened at Scott AFB on the way home. We requested a practice no gyro PAR (Precision Approach Radar) approach. These are the easiest and safest approaches in bad situations especially if you are single pilot. The controller aurally flies your aircraft down the localizer and glide slope using your radar return. All you do is fly basic instruments and follow directions. The directions are simple as well: turn right or turn left or stop turn or begin climb or adjust climb or begin descent or adjust descent or level off. Once you have been guided to the localizer outside the outer marker and are turned inbound you will be turned over to a final controller who will instruct you not to respond to further directions but to continue the approach on the ILS if no instructions are heard for five seconds. This is standard practice allowing the controller to have full and immediate use of the radio for rapid corrections on the final phase of the approach. A normal no gyro PAR would sound something like this: "Guard 191 is slightly left of course turn right. Guard 191 stop turn. Guard 191 turn left. Guard 191 stop turn. Guard 191 is on course two miles from the outer marker. Guard 191 on course one mile from the outer marker. Guard 191 is at the marker, begin descent, I need 432 feet per minute (the rate of descent is in the IFR Supplement but given as a courtesy). Guard 191 going below glide slope, adjust descent. Guard 191 is left of course, turn right. Guard

CONTACT FLYING

191 stop turn. Guard 191 is slightly below glide slope adjusting nicely. Guard 191 is slightly left of course adjusting nicely, on glide slope." This goes on until you are asked to take over visually at decision height. Controller turn and climb/descend commands are continued, however, until you call missed approach or field in sight.

Shortly after approving our request for the no gyro PAR, we were handed off to a nervous sounding female final controller. While inbound from beyond the localizer she commanded us not to respond to further instructions. The next command was, "Guard one niner one turn right." After a few minutes and three complete 360 degree right turns, she hurriedly commanded, "Guard 191 stop turn!" She had either forgotten us or frozen up. Of course, after five seconds, we should have continued on the ILS using the CDI but Joe was always looking for an interesting situation. On the ground while visiting Base Operations and the Consolidated Dining Facility Joe asked every female airman we encountered, "Pardon me Airman, are you the final controller?" in his very dry and serious manner. He didn't get slapped but he had the Major and me rolling.

Flying in the Missouri Guard taught me a lot about instrument flying. Many approaches were made to minimums in extensive stationary frontal conditions. Somehow we managed to find something that would qualify as an alternate but we never intended to go there. With only 2.5 hours of fuel, few options are available in hard instrument weather. We simply slowed down to 40 knots and made the destination. I have no use for single pilot instrument flight especially now that I require short and medium range readers. Much greater safety can be accomplished with two sets of eyes.

Army regulations dictate that all cross country flights will be filed IFR (Instrument Flight Rules) unless the mission requires VMC (Visual Meteorological Conditions). Due to Midwest weather and no oxygen on Army helicopters, most cross country flights and Annual Instrument

Proficiency Checks were conducted during IMC. CW3 Mike Henry, who is now head of training at the FAA in Washington, gave me my Annual Instrument Proficiency Check in 1973. We were holding over the LOM (Locator Outer Marker) at Richards-Gebaur AFB in Kansas City. We were in the clouds with two aircraft holding above us and one aircraft holding below us on the same LOM. Mike pulled the alternating current circuit breaker out leaving me with needle, ball, and airspeed. Army turbine helicopters had no vacuum system. The rate of turn needle was powered by direct current from the 28 volt nickel cadmium battery. Most of my 300 hours of actual instrument time was in Army helicopters in Missouri National Guard. Midwestern Air and Army National Guard units have, by necessity, very good instrument instructors.

I have made several actual instrument approaches into March AFB and Norton AFB in the Los Angeles Basin. From 1975 to 1983 I flew as a medical evacuation pilot, section leader, and flight instructor with the 717th Medical Detachment (Helicopter Ambulance) in the New Mexico Army National Guard out of Santa Fe. To complete summer camps in 1979 and again in 1982, we provided the actual medical evacuation for large regular Army field exercises at Ft. Irwin in the Barstow-Daggett area of eastern California. One of our Hueys (3rd Up) was always available to back-haul wounded soldiers from the jet-pack blow-up Army Reserve Combat Support Hospital to larger hospitals in Los Angeles. Our regular injuries went to March AFB and our burn patients went to Loma Linda civilian hospital.

The 717th provided six Hueys (UH1-H) on the Ft. Irwin missions. Three helicopters were always immediately available: "1st Up" was on the pad ready to crank and depart within one minute. This was possible because it had already been preflight checked, run up, and everything left on except fuel and master switches. Even the linear actuator (to beep the engine slowly up to full RPM after the throttle had been fully advanced) was left in the beeped up position.

CONTACT FLYING

1st Up crews slept in flight gear at the ready tent. When a mission came in the AC (Aircraft Commander) went to the TOC (Tactical Operations Center) to get the grid coordinates and any known information on the extent of injuries. The pilot, crew chief, and medic (all Army crews cross-train to provide medical backup) ran to the helicopter and cranked. The pilot turned on the battery and fuel and hit the start trigger. As soon as the AC ran from the TOC and jumped into the left seat, the helicopter lifted off. The AC pointed in the general direction of the location of the injuries, held up fingers to indicate to the medic how many injured soldiers would be treated, strapped in and put his helmet on before hooking into the intercom. By this time the pilot had flown halfway to the pickup point. Our average time was ten minutes from receiving a mission to completing a mission at the Combat Support Hospital. Lots of adrenaline! Lots of fun! It was dangerous but reasonably safe when everyone was totally involved. We were a good unit.

There is a safety switch for sensitive uses on military aircraft that is spring loaded to the **ON** position. The only way to get it to the Off position is to pull the switch up and across a detent. This switch, and the saying derived from it, best describes the intensity and joy of high adrenaline work like air cavalry and medical evacuation. Truly we were "spring loaded to the On."

The job of the 2nd Up crew was not as intense. When 1st Up left, the 2nd Up crew dressed and became 1st Up until 1st Up returned or called back up when near the completion of the active mission.

The gentleman's work was 3rd Up. You slept in your jammies and made normal preflight preparations for scheduled back hauls to established military or civilian hospitals. This is where we got many actual instrument approaches into the Los Angeles Basin. Actually it was a milk run. We used the Evac call sign which gave us the highest priority after the President. All of our approaches followed vectors direct to the LOM. Also in those days most of us used the PAR. Because of the high training and manning requirements, many Air Force bases no longer provide the PAR. This is unfortunate because it was the safest approach available and required the only basic instrumentation in the aircraft.

CW2 Dave Hawkins and I were flying a Huey IMC to Joplin and back to our unit, the AVCRAD (Aviation Classification and Repair Depot) at Springfield, Missouri. I was the Cobra Maintenance Test Pilot at the AVCRAD when Missouri only had one Cobra in the whole state. I never flew it. Anyway on our return from Joplin the controller at Kansas City Center asked us why we were 12 miles north of course. We were showing a centered CDI. VOR failure does not mean the ILS (Instrument Landing System) receiver has failed but it makes you wonder.

Just before arriving at Springfield VOR, the IAF (Initial Approach Fix) for the ILS front course, we were vectored for the ILS back course. My mind was fried by the time I got onto the localizer and I was positive rather than reverse sensing the CDI. Dave said, "back course" a couple of times to try to reorient me but I was confused. I asked, "Dave do you know where we are?" He answered in the affirmative. I said, "you have the controls. Fly the approach." Safety wise, there is no substitute for two pilots.

Instrument flying is clean, precise, highly regulated, by the numbers, legal, left brained, logical, sequential, programmable and boring. You can get caught up in the goodies, the power, and the assurance of it, but it just isn't as much fun as contact flying. It requires organization and work. If you like it, the present primary training curriculum well prepares you for it. Be careful however. Much of what you do in your C-152 simulates what is done in a Piper Aztec. You are not in a Piper Aztec, however, and you must get into this highly controlled instrument environment without hitting obstructions in your C-152. To accomplish this safely you need contact flying skills.

Chapter 1

STALLS AND SLOW FLIGHT

In the early 60s I was practicing chandelles in a Cessna 150 east of Montrose, Colorado. I was finding that there was no altitude gain with the maneuver and that in fact I often lost altitude in the mush down during the slower part of the maneuver. Then I inadvertently spun out of the slower, supposedly altitude gaining, part of the maneuver. I began to question maneuvers in general and whether pitching up ever actually meant climbing in a small aircraft.

The text of this contact flying manual starts with slow flight and stalls. This is awkward for me because I actually first understood stalls and slow flight when the Army taught me the brisk walk apparent rate of closure technique to land helicopters. As my instructional technique evolved with the needs of my students, I came to the point some years ago where I taught landing first and stalls later because the stall is best learned from the contact (no instruments) landing. However, since stalls are taught first in a normal instructional program and high altitude stalls are practical test requirements, I will start with them. They will make more complete sense after you have read chapter five on the brisk walk apparent rate of closure approach. If at the end of this chapter you are confused, read on through chapter five then return to this chapter.

We need to be able to control the aircraft in slow flight because we must slow the aircraft sufficiently to land. We must stall the aircraft to land with minimal loads against the gear, especially the nose gear in tricycle geared aircraft.

Landing is a ground reference maneuver. Why then must we teach stalls and slow flight at altitude with little or no reference to the ground and completely divorced from the landing situation? With the incredibly long runways available to us today why not teach stalls and slow flight on these runways? At Contact Flying and Aircraft we concentrate, pre solo, only on those things required to take off and land an airplane. This includes stalls as required by 61.87 (d) (10). However we teach more stalls and slow flight in conjunction with landing than at altitude. We feel we teach a better understanding of slow flight and stalls with the contact method. We also guarantee solo prior to reaching the first learning plateau. Delayed solo in regular flight schools results in a significant number of dropouts.

The AOPA Air Safety Foundation studied more than 2,000 instructional accidents, both dual and solo, from 1992 to 2001. In that period 42 or 29.8% of dual fatalities (# of accidents) and 12 or 20% of solo fatalities occurred during maneuvering flight. I believe we are loosing too many good instructors and students practicing maneuvers to such close practical test standards that instruments must be used to meet those standards. Instruments portray data slowly and incompletely through only one of our senses. This pales in comparison with the immediate and complete data contact pilots receive by ground reference, kinetics including stick position, aural information, and smell.

The problem with stalls, as presently taught, is that they develop the mind set that such nose high attitudes at minimum controllable airspeed are quite normal. They are not normal except on landing. The great amount of time devoted to stall practice at altitude reinforces the deception that this is a normal attitude for takeoff. At Contact Flying and Aircraft we emphasize that the stall is a useful maneuver in landing and is neither safe nor necessary anywhere else. Yet during the short field takeoff over an obstacle the PTS dictates that we climb at a high pitch attitude near the ground. This is neither normal nor necessary and it is dangerous. The acceptance of such an unusual attitude on takeoff and climb out is deceiving and dangerous. Little airplanes haven't the necessary power to climb at a high pitch attitude and therefore should not be forced to do so.

CONTACT FLYING

If Private and Commercial Practical Test Standards force the student to climb out at high pitch attitudes to perform the short field takeoff, he may assume that this is normal in small aircraft. LSA Practical Test Standards allow the airplane to be flown level in ground effect if common sense and our five natural senses deem it necessary. Small aircraft will not climb at Vx for very long. The physics of Vy tell us the airplane really dislikes Vx. If the student persists in holding the Vx pitch attitude until energy bleeds off, a mush then finally a stall will result. Climb out is the wrong time to practice a stall because of the limited altitude. Yet high altitude stall practice doesn't give enough ground reference to fully understand slow flight and stalls.

Why not teach slow flight and stalls in ground effect? At Contact Flying and Aircraft we teach stalls safely on the mile long runway at Monett, Missouri. A five foot altitude over this long runway gives adequate ground reference to fully understand the high angle of attack. Slow flight (airplane hover taxi), mush down, and stalls at an altitude of five feet over this long runway help us keep the stall in its proper perspective which is to slow the aircraft sufficiently for landing.

Slow flight, mushing down, and stalling can be taught over long runways. Multiple stalls can be taught on one circuit. On takeoff we can demonstrate and allow the student to observe the error of trying to climb when the aircraft has insufficient zoom reserve in the form of airspeed.

To sustain flight we need energy (zoom reserve) to move the wing forward with enough speed to create enough lift to overpower the total weight of the aircraft. If we have altitude we can use gravity to achieve this necessary speed. If we stay in ground effect Newton's Law that "every action has an equal and opposite reaction" helps sustain sufficient lift to make sufficient speed quickly achievable. The air forced downward by the curvature of the upper portion of the wing will impact the earth and cause an equal amount of reaction in the form of buoyancy. Small airplane cruise speed can be achieved in ground effect even on short runways. This achievement of what is called zoom reserve in the form of airspeed makes maneuvering flight possible within the confines of a short strip. When we have zoom reserve in the form of airspeed we are buoyant and we can maneuver the aircraft safely.

Yes, when we have zoom reserve in the form of altitude we can use gravity to trade that altitude for airspeed. When we attempt Vx on the short field takeoff, however, we not only put ourselves out of the benefit of ground effect but we also set the airplane up for a mush into the obstacle or stall above it. Physically we can trade the altitude, gained with the high pitch up, for airspeed when we hit a downdraft. The fact that students and even experienced pilots have trouble with this is born out in accident statistics. A very hard push forward on the stick just as the obstacle comes into serious view may be required to maintain sufficient relative wind to prevent a mush or stall.

Airspeed and altitude are two forms of the same energy which Wolfgang Langewiesche calls "zoom reserve" in <u>Stick and Rudder.</u> He also states accurately that, "airspeed is altitude and altitude is airspeed." But on takeoff, he says, we prefer to have our energy or zoom reserve in the form of airspeed. Attempting to achieve zoom reserve initially in the form of altitude just does not work with low powered aircraft. By pitching up to Vx or even Vy in gusty conditions we load the aircraft more than it can handle efficiently. The result is a mush or stall. Private and Commercial Practical Test Standards stipulate a high pitch attitude takeoff for a normal takeoff and a very high pitch attitude takeoff for a short field takeoff over an obstacle. This puts our students at great risk for a takeoff and departure stall. They have difficulty pitching down significantly, trading altitude for airspeed, when confronted with a downwind shear or downdraft. This hard pitch down is bound to be late if we are using a trend instrument (airspeed indicator) to indicate when to pitch down.

Because the PTS demands it, we must teach the short field takeoff to those trend instrument (airspeed indicator) standards. However, it is not disqualifying to recover from an imminent stall, especially while in the pattern. Thus I am sensitive to the slightest glitch in aircraft performance when teaching this dangerous maneuver. And furthermore, I am not depending on a trend instrument to indicate that there is a problem. Any manufactured light aircraft will tell you it is hurting and that continued lift is marginal at present angle of attack long before the airspeed indicator winds down to that conclusion. Also I teach a lot about how the aircraft acts and about its preferences during this critical phase of flight.

Contact Flying and Aircraft uses long runways to safely simulate the downwind shear problem. Immediately upon pitch up we pull a little power off to demonstrate a downwind sheer. At the same instant we instruct the student to recover from the resultant mush by leveling the nose and recovering in ground effect flight. We emphasize throughout this maneuver that the wings be kept level and the nose aligned with the centerline using rudder only. The use of ailerons close to the ground is unproductive and dangerous. We also emphasize that the problem can be completely avoided by staying in ground effect on all takeoffs.

Stalls are effectively taught when complying with the PTS soft field landing procedure. The soft field landing is a long powered hold off. Slow flight can be taught as well. Just before touchdown on the numbers, we require the student to add power and fly the aircraft in ground effect well down a long runway before allowing the aircraft to touchdown by reducing power. As with takeoff, we must emphasize the use of rudder and not aileron near the ground. If there is a steady crosswind the wing must be set to counter drift and then left alone. Only if the crosswind is gusty will aileron changes be necessary. The student must appreciate that in the sloppiness of slow flight the aileron is his enemy, not his friend. Overuse of aileron is a big problem for students on approach and landing. As the aircraft slows all control response except throttle diminishes. From a pilot perspective however, pitch and yaw rates seem less diminished than does roll rate. This throws our timing off.

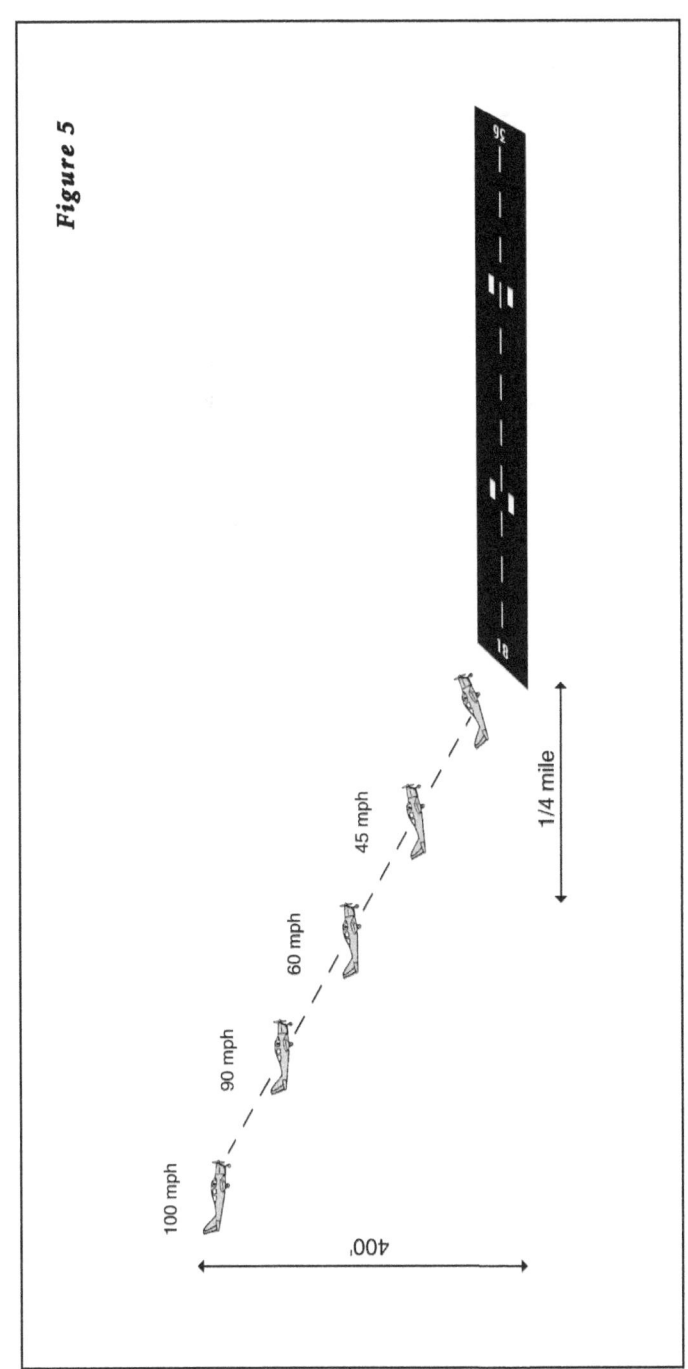

Figure 5

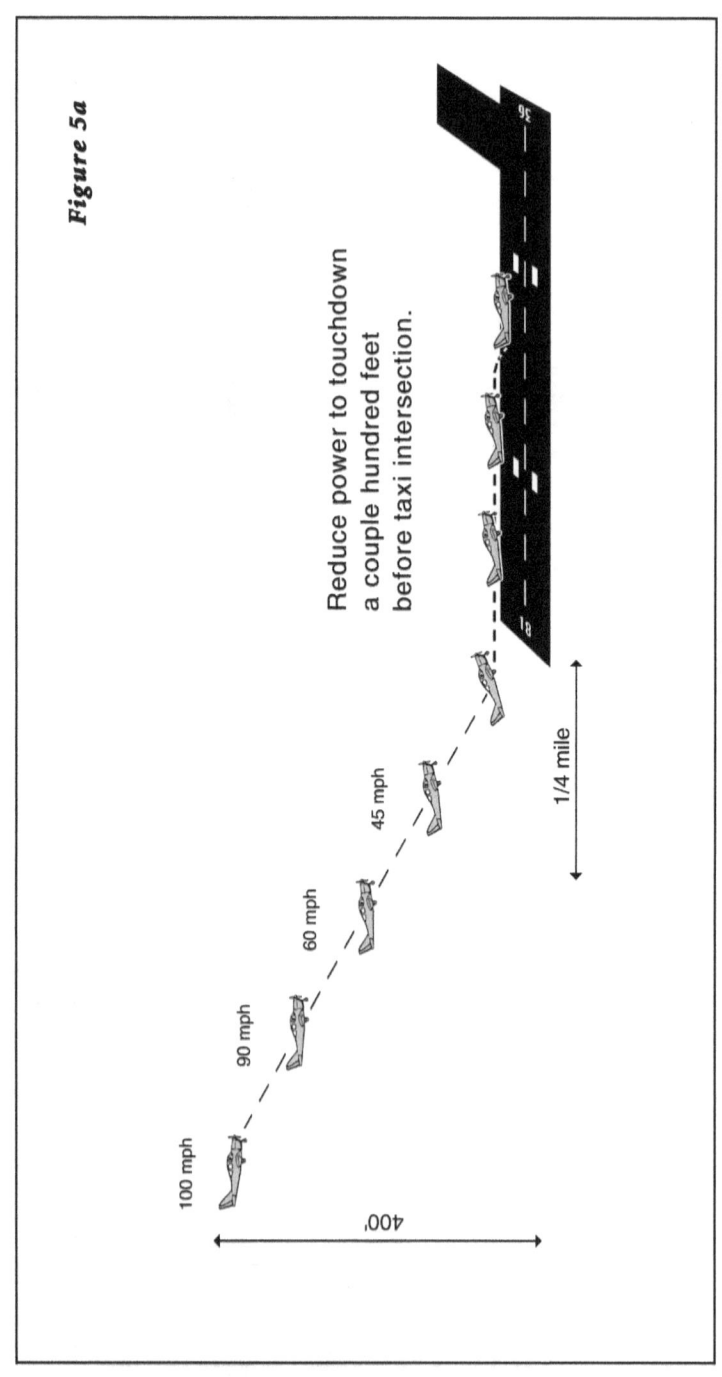

Figure 5a

Once aligned with the runway (on final), yaw control (rudder) is sufficient to keep the aircraft aligned with the runway. We use rudder not aileron to turn the aircraft on or near the ground.

When the wind is strong, steady and directly down a long runway a helicopter like hover taxi is possible with fixed wing aircraft. *See Figure 5a*. This technique allows excellent slow flight and stall training and it teaches the student that the wind can be our friend. We have the student set up a normal brisk walk apparent rate of closure approach with full flaps. At the point where he levels the aircraft just prior to touchdown for a wheel landing or just before touchdown for a tricycle full stall landing, we have him use power as necessary to maintain one to five feet of altitude over the runway while maintaining the brisk walk apparent rate of closure. Because the groundspeed is slow, the one to five feet of altitude may be held until near the last taxiway before power is reduced to effect touchdown under ample speed control to make the turnoff. What I mean by ample speed control is very near stall speed. A small aircraft can fly perfectly well and safely below stall speed if in low ground effect (1" to 5' AGL) using the brisk walk apparent rate of closure procedure. We point out the high airspeed to the student only to illustrate the primacy of what is happening outside the aircraft over what is happening inside the aircraft (airspeed will be high but groundspeed will be no different than a normal apparent brisk walk.) There is no way our Pilots Operating Handbook can give us numbers that will work as effectively as looking outside the window. It is impossible to perform this maneuver using instruments. We can hover-taxi a fixed wing aircraft only by reference to the ground.

At Contact Flying and Aircraft we teach the art of flying using ground reference rather than instruments while near the ground. This is entirely possible and desirable. Moreover it is extremely necessary in less than perfect conditions. Mastering this concept breeds "real" confidence.

CONTACT FLYING

Our advanced students are able to hover-taxi a small aircraft in a no wind, crosswind, or light downwind condition. Now the airspeed will be the same as in the upwind or no wind hover-taxi but the groundspeed will be higher and care will have to be taken not to either touch down early or gain too much groundspeed. We teach the hover-taxi down long runways until 500' short of the turnoff taxiway.

The brisk walk apparent rate of closure approach is an Army helicopter training technique that stabilizes an approach based upon the relative appearance of things on the ground rather than upon numbers on gauges. It uses Mr. Langewiesche's "stall down" technique. By keeping the runway number or any object in the desired touchdown zone closing with us or coming toward us at a steady apparent brisk walk, we will arrive at the touchdown point with no zoom reserve either in airspeed or altitude. The beauty of this approach is that we have zoom reserve (safe airspeed and altitude) out over the final approach obstructions, yet we lose both reserve airspeed and reserve altitude by the time we need to touch down on the numbers. The optical use of the apparent fast walk closure rate causes a continuous decent and deceleration to touchdown. *See Figure 5.* This results in a spot landing every time.

With higher obstacles on short final the glide slope angle will be greater but the brisk walk apparent rate of closure will be the same. *See Figure 5a.* This landing procedure is simply continued down the runway in ground effect if we wish to practice slow flight and controlled stalls. By using the hover-taxi until just short of the desired turnoff intersection, we naturally and usefully practice slow flight and a stall each time we land at a large airport and hover-taxi to the desired intersection.

High altitude stall practice does not transfer well to contact flying because there is no accurate visual reference to the ground and because instruments are used to establish parameters for causing the stall and for the recovery. Instruments, with the exception of the VSI, show the student little about what is happening during the mush before the stall and the actual stall. At one foot above ground level the student can see that a mush will put him on the deck as easily as a stall. At one foot above ground level the student can discriminate altitude to within plus or minus one foot. At one foot above ground level the student can see the error in using aileron to maintain wings level and runway alignment. Slow flight and stall practice in ground effect over long runways will transfer well to takeoff and landing situations and will give the student great confidence in his ability to handle the aircraft in these ground reference operations. Takeoff and landing are ground reference maneuvers. No instrument should be necessary to perform them in small aircraft.

In helicopters the maneuver that resembles the mush and stall is settling with power. Settling with power can occur on pull up from a spray run or gun run or while hovering in a no wind or downwind condition. It results from insufficient power to fly in ones own rotor wash. In airplane terms this would be called exceeding the critical angle of attack. The solution is always to reduce the angle of attack of the tip path plane (think big round wing) by lowering the nose with cyclic or reducing pitch in all blades by lowering the collective or by both pitching down and lowering the collective. In airplane terms we would lower the nose and reduce power. While reducing power might be incorrect for airplanes, the principles here are the same. We need to return to an angle of attack and relative wind situation acceptable to the aircraft. Unfortunately, in helicopters, the aircraft will not tell us what it wants to do. Even small helicopters, like large airplanes, fly very artificially and very dangerously.

It is unrealistic to expect that my ideas on the apparent rate of closure approach or slow flight and stall practice over long runways will ever become the norm. It is hard for an old dog to learn new tricks. With old pilots we at Contact Flying and Aircraft teach contact flying to increase skill levels rather than to add licenses, certificates, and titles of nobility. It will improve an old pilot's confidence and ability. It will also make him a safer and more versatile pilot. The satisfaction has to be internal for it cannot be certified.

CONTACT FLYING

Young zero time and pre-solo pilots are another story. Ladies and gentlemen, this old bold pilot can teach you the art of flying in the really old style quicker and easier than your golf pro can teach you to hit a ball straight. Unless your mind has been polluted with a lot of numbers, contact flying is easier and safer than driving a car. Thankfully light sport flying regulations allow fast and effective training through solo. Private training programs which overload students with infinite knowledge and performance standards often delay solo well beyond the tenth hour. This solo after reaching the first learning plateau leads to self doubt, incrimination, and poor flying techniques. We old, bold pilots are shocked by the higher number of hours commonly required to solo and the obviously resultant lack of confidence among flight students today. Do onerous regulations, test standards, and legal considerations cause many students to peak and burn out prior to solo? Students don't need to perfect all manner of stalls to solo safely. Thankfully the LSA Practical Test Standards reduces the number and types of stalls required.

Chapter 2

THE TAKEOFF

Early in my flying career I got very close to a wire at the end of a grass airport at Marshfield, Missouri. When strong south winds dictated a takeoff south one had to get airborne on the down slope of the north hill, fly across the valley, touchdown and again takeoff on the upslope of the southern hill and clear a telephone wire on the south end. I tried to maintain too much altitude through the valley but seemed to get off the southern upslope in good shape. My problem was that I tried to miss the wire by too great a margin and almost mushed into it. The feeling that I lacked control of the aircraft during the short field takeoff with an obstacle caused me to lessen the altitude margin.

While frustrated with the accepted short field takeoff procedure, I continued to use it until conditions dictated otherwise in the Rocky Mountains. I was the instructor and scheduling officer for Tohatchi and Gallup Squadrons of New Mexico Civil Air Patrol. Both airports were at 6500 feet above sea level. At common density altitudes in the Rockies Vx does not exist. Even Vy is too great an angle of attack with common afternoon winds of 30 gusting to 60. Therefore I began teaching a soft field type takeoff and staying in ground effect until cruise speed was achieved out over the desert. This speed allowed safe maneuvering to miss terrain while taking advantage of hydraulic and thermal lift.

To understand the takeoff and its problems, we need to understand why we takeoff. That is obvious you say; we simply want to get airborne. And you are correct. Practically speaking, the objective is to get airborne and operate in the fluid of the air rather than with the restrictions of land and sea navigation. But when we consider the bureaucratic burdens of Federal

CONTACT FLYING

Air Regulations and Practical Test Standards we find rigid speed and altitude standards that may infringe on common sense and even safety.

What then is our objective in the takeoff? Is it to get up to high altitude (above five hundred feet AGL) as soon as possible or is it to secure buoyancy, firm control feel, good relative wind sounds, and maneuverability (what is called zoom reserve in the form of airspeed) as soon as possible? There is a very helpful chapter in Stick and Rudder, by Wolfgang Langewieche that explains the relationship between airspeed and altitude. According to Mr. Langewieche, "airspeed is altitude and altitude is airspeed" and during takeoff out of a tight field we prefer to take our energy, or "zoom" as he calls it, in the form of airspeed.

For contact flyers climb winds, the hemispheric rule, V speeds, altitudes, or even climbing out are useless concepts. Going up eats time and energy, heats our engine, degrades our maneuvering speed, and prevents good visual reconnaissance. Our objective is to achieve zoom reserve in the form of airspeed as soon as possible and thereby become a mission capable air machine. In small, low powered aircraft, it is often dangerous to attempt altitude over airspeed especially when we use arbitrary speeds to achieve that altitude.

Flight training is not just a study of fact. It is also a study of belief. If we think students leave the training phase of their flight careers with a vast database of aviation facts which can be applied at a later date to all aviation situations, we are sadly miss-oriented in our teaching. We are training people not computers. When things are not going according to standards and the aircraft is mushing down, many pilots will pull back on the stick because they *believe* this will cause them to stop going down and even go up as they wish. At Contact Flying and Aircraft we train students to believe airplanes will not climb continuously. At least the airplanes most of us fly will not climb continuously. In fact there are times when pulling the stick back will kill us. Airplanes are designed to prefer flat flying. We train our students to believe airplanes prefer flat flying. They come to believe continuous climbing flight will eat their zoom reserve (first airspeed, then altitude) and send them crashing toward earth no matter what. They learn to believe in zooming or thermaling or finding hydraulic lift rather than in a stabilized climb at some V speed. Many pilots believe that altitude is safety and that it is gained by pulling back on the stick. Our students believe airspeed, buoyancy, and low angle of attack are safer because maneuvering flight is thereby possible.

There are too many illusions in a flight training program that insists on V speeds that can be read from gauges. A real killer illusion is Vx. Here is a hard, factual number published in the Pilots Operating Handbook for every modern aircraft. Its arbitrary use creates the illusion that small aircraft can climb forever at this magic number, constantly gaining the most altitude possible over distance without regard to other factors like density altitude, load, updrafts, downdrafts, wind shear, and constantly changing angles of attack based upon variables not controlled by aircraft speed. And that speed (Vx), which many pilots believe will gain them altitude, is based on the most unreliable of sensory cues in contact flight, an instrument inside the aircraft. Furthermore, the very instrument that gives many pilots *the fact* upon which they are betting their life is known to mislead on a regular basis.

The airspeed indicator will give our contact student, and instrument students as well, no reliable fact as to the current speed of the aircraft. It gives him a trend, an indication of what the aircraft has done in the past and some prediction of what it might be going to do in the future (based on that trend). It certainly does not tell him that he has lost adequate buoyancy or the zoom reserve necessary to safely maneuver the aircraft as necessary to miss obstructions.

There are many clues, not found on the panel, that will give our contact student reliable facts about the speed of the aircraft across the ground, where exactly (between trees, under wires, etc.) it is going (not heading),

CONTACT FLYING

and what kind and amount of energy is available to him in the accomplishment of his mission.

While airspeed is so unreliable in contact flight where many other sensory perceptions are more available and more accurate, airspeed is very crucial and reliable as a trend setter in instrument flight. Through generalization and integration it has gained undue prominence in the primary flight program. At Contact Flying and Aircraft we believe it is very dangerous to give airspeed such prominence over visual, aural, sensory, and kinetic clues that give us a sense of angle of attack, buoyancy, and what the aircraft wants to do.

In the takeoff phase of every flight the main concern is getting airborne and clearing any departure obstructions. With airplanes we have to consider the surface of the runway. Roughness in the first three hundred feet will slow our acceleration, but when we have accelerated to partial buoyancy, a rough section of the field might help bounce us off into ground effect flight. It is a good habit to get off the surface (with flaps if we have them) and into ground effect as soon as possible. Once off the surface and in ground effect, the remaining takeoff surface condition is of little consequence. We should accelerate in ground effect through stall speed, through Vx, through Vy, to a reasonable zoom reserve in the form of airspeed (we acquire this comfortable feeling through all our senses) prior to leaving ground effect. To leave ground effect sooner than absolutely necessary is unwise because of our poor engine power in comparison with the power of temperature, wind shear, humidity, density altitude, and gravity. Natural laws have more power than our engine.

The slope or grade of the runway is important to both airplane and helicopter operations. Any runway with greater than six percent grade (the steepest grade on major highways) will warrant a downhill takeoff regardless of wind direction.

Tricycle airplanes are at a great disadvantage in comparison with tail wheel airplanes and helicopters on any surface that will allow the nose wheel to mire. On a surface that a car could easily be driven across, we find our trike stopped early in taxi or takeoff by a sunken nose wheel and perhaps a bent prop. Any hindrance of the free rolling of the main wheels (snow, heavy vegetation, ice or mud) will throw the whole weight of the aircraft onto the nose wheel. What happens to the nose when the brakes are applied? Yes, it goes down compressing the nose gear strut. On a boggy surface braking will bury the nose wheel. This is why tail wheel airplanes with poor brakes were conventional in the first place. This problem can be overcome if sufficient power is available to raise the nose wheel by application of full power and full up elevator from the beginning of taxi until the nose wheel lifts free of the surface (remember, we steer with the rudder). As the nose comes up we can relax the throttle enough to allow the nose wheel to just clear the top of the snow, grass, or mud. On takeoff we relax the elevator to keep the nose wheel just off the surface.

We must consider the effect of runway length, runway condition, grass length, moisture content, and density altitude on the ground run. We could refer to the POH (Pilot's Operating Handbook) for charts, graphs, and sliding scales. The POH will help one gain light load experience safely but spraying, bush operations, gunship operations, and troop lift operations require heavier load experience than what is allowed by the POH and regulations. Gaining heavy load experience in these operations requires moving from the POH and regulatory limitations to greater than maximum gross weight loads.

Allowance is made for this in Alaska where aircraft may legally be loaded to one hundred fifteen percent of gross weight. What, from a safety standpoint, is the difference between Alaska and Cheyenne, Wyoming? How can a safety board rule that the crash of the Cardinal with the budding young pilot in Wyoming was caused by overloading when the same load would not be considered the cause of the crash in Alaska at the same altitude under the same conditions? Can governmental double talk

CONTACT FLYING

be the cause of a crash resulting from improper training? In the interest of safety, Contact Flying and Aircraft will train you for the real situation rather than the situation as described by regulatory double talk.

The takeoff crash of the Cardinal with the young girl in Cheyenne was caused by the mushing that resulted from trying to climb when conditions would not allow climbing. We all saw the news clip of the Cardinal going by at a very high pitch attitude but not climbing. The instructor flying the plane was taught the importance of climbing to within four hundred feet of an arbitrarily determined pattern altitude before turning. Best angle of climb speed was closer to best rate of climb speed at that density altitude but whether he attempted Vx or Vy is moot. In this real life situation (which may never have occurred in his training and experience) the important thing was not to leave free, natural energy on the airport. He should have elected to take off toward descending terrain using the universal law of gravity. More importantly, he should have stayed in ground effect using the Creator's law, which Newton discovered, of equal and opposite reaction until he felt he could comfortably zoom over, or rudder turn around, any obstructions preventing his getting to lower ground. Yes, slightly rising terrain was a factor in the crash but the nearness of a thunderstorm and local winds from 280 degrees at 20 gusting to 30 knots (airport winds were much lighter) need not have been. Properly trained contact pilots learn to use the tremendous force of high winds over mountainous terrain to hydraulically lift their little aircraft at rates of climb unattainable anywhere with engine alone. In similar conditions survival is possible. It is even routine in the mountains in slightly overloaded (in the lower 48 states only) aircraft and with smaller engines than the Cardinal.

The instructor might not have crashed had he been trained by an old crop duster. He would have been taught to keep the nose down and the speed up at critical times like takeoff and to expect the unexpected at all times. He would have stayed in ground effect until he had developed zoom reserve in the form of airspeed even if this meant flying between the isolated houses on the five acre lots around Cheyenne. He would have found hydraulic lift over the rising terrain and then turned sharply allowing his nose to fall naturally into the valley associated with the ridge he had gotten the hydraulic lift from. He would have followed the valley, in ground effect if necessary, to the lower terrain east of Cheyenne. Nothing in the normal training curriculum or the POH prepared him for what he encountered on a fairly normal morning in the high plains. But if he had to make the flight (he could have stayed put), contact skills were needed. They are needed because VMC takeoff anywhere is a ground reference maneuver, not an instrument maneuver.

Contact operations, which include takeoff and landing, cannot be learned solely from books, charts, graphs, observance of Practical Test Standards, and regulations. They can be learned only from experience, good and bad. Under present regulations a qualified instructor will be helpful only if that instructor has had real life experiences. These life, and near death, experiences are not enumerated in logbooks nor are they tested for by present written and flight tests. Contact Flying and Aircraft offers a curriculum where contact flying is required through solo. We feel this greatly benefits pilots in general and raises the safety bar considerably.

Getting the ship off the ground in ground effect and out over the crop or desert, while remaining in ground effect, is only half the takeoff operation. Now we must either zoom over or miss laterally any obstruction in our flight path. So we come back to the airspeed over altitude controversy. We who operate heavy at low altitude simply cannot survive using the Vx solution to obstructions either on takeoff or in the field. Our school teaches that this technique is unsafe for anyone flying low powered or maximum loaded aircraft. *At high density altitude or when loaded to gross weight, attempting either Vx or Vy prior to achieving zoom reserve in ground effect can and often does result in mushing or stalling.* We believe that attempting a Vx takeoff or go around and then responding, in a slow and stabilized manner, to airspeed trends accounts for a sizable share of our fatal takeoff and go around accidents.

CONTACT FLYING

As a pilot or instructor taking off from a short field with obstacles at the end, have you ever experienced a definite sinking feeling accompanied by a more labored engine sound, less relative wind noise, worse visibility over the nose, less positive resistance on the stick or yoke (which is well back), the smell of gasoline, and the sound of the stall warning devise? This all too common situation during takeoffs using Practical Test Standards criteria results in too many takeoff and departure stalls being practiced in actual conditions.

Have you ever watched a small crop duster work a field with obstructions at the end? Does this little 235 horsepower Pawnee or 260 horsepower Cessna appear awkward over those obstructions? Those old spray planes are similar, in power to weight ratio, to small people airplanes. The spray pilot's angle of attack is greater (but he doesn't try to maintain it) and his bank angles would fail him on any practical test but he doesn't insist on level, high load factor turns. You may have believed the crop duster could engage in such amazing maneuvering flight because of greater engine power but that is not the case. The engine is not powerful. He is managing powerful natural energy. Both the spray plane and the small airplane are at full throttle with low power to weight ratios. The difference is that the spray pilot is using ground effect to develop zoom reserve in the form of airspeed, airspeed to develop limited altitude (never too much is asked), and finally that developed altitude is traded for airspeed as he allows the nose to fall through naturally. Never does he allow himself to fall into the trap of the level, high load factor turn.

Are you, as am I, shaken and terrified by the deck angle of airliners on takeoff? For those of us with thousands of hours in light aircraft, any rocket powered takeoff is surreal and troubling. These large and powerful aircraft make useful application of those V speeds examined on the Private Pilot Practical Test. This is the ultimate in integration of contact and instrument skills to arrive at a seamless transition to larger aircraft.

And your FAA did not entirely hang you out to dry in your 152. They know it does not fly like a jet. They, like me, use exaggeration as a training devise. But they attempt to protect you through regulations and a sanction oriented culture. In this culture "get his number" is entirely appropriate and useful. The fear of sanctions limits the number of pilots willing to chance maneuvering the aircraft in close proximity to the ground. Unfortunately these same sanctions, and fear of them, limit the ability of pilots to learn to maneuver their aircraft in close proximity to the ground. This book and Contact Flying and Aircraft attempt to show you the other side of the coin so that if you choose to investigate true contact flying, without instrument integration limitations, you may become a more rounded pilot.

On takeoff over a 50' obstacle, the PTS implied objective is to achieve altitude as soon as possible and go over the obstacle as high as possible. If the PTS would just combine the soft field takeoff with the short field takeoff, much safety could be achieved. We at Contact Flying and Aircraft require our students to lower the nose to stay in ground effect until at least Vy (I don't really advocate an arbitrary speed on a trend gauge). What about the 50' obstacle you ask? Try both techniques on the same field in the same conditions with your airplane. Compare the soft field takeoff and staying in ground effect until Vy technique with the pitching up to a Vx attitude slightly before achieving Vx (airspeed is a trend instrument, you have to assume you are at Vx) technique. I guarantee you that the soft field takeoff technique will put you over the obstacle with the outcome of the maneuver far less in doubt.

On too many takeoffs over obstacles, whether practice or real, neither the student nor his instructor have positive control of the aircraft and the instructor does not have a positive feeling about the outcome of the maneuver. Students with any sense of what the airplane wants to do are uneasy with the PTS short field takeoff. Using what is taught, based on the Practical Test Standards, we seldom achieve what Wolfgang Langewiesche calls "zoom reserve" and I call zoom reserve in the form of airspeed. As

Mr. Langewiesche puts it on page 85 of <u>Stick and Rudder</u>, "The inexperienced pilot has a strong tendency to point the nose up steeply and simply hope that the airplane will climb out. But it is easy to overdo this and, by trying to get too much altitude too quickly, to kill one's speed and stall, perhaps 50 feet off the ground."

In today's terminology, the PTS correct pilot has a strong tendency to point the nose up to Vx (pitch attitude) and simply hope that the airplane will accelerate to Vx and maintain Vx. But it is easy for him to properly do this and fail to realize he is sinking at Vx or below due to conditions outside the PTS. His instrument recovery from the resulting stall or imminent stall does not put him as high over the tree as a good soft field takeoff would have. This problem is evident but not cited in many recent takeoff accidents. The strict adherence to V speeds plus or minus a small margin, leads to standardization in technique and fairness in flight testing. They do not lead to safety.

Below are some accident reports from the NTSB database. Based on statements of pilots in takeoff accidents in the last ten years, it appears pilots leave regular training programs with a lack of gravity, ground effect, buoyancy, shear, and zoom awareness (italics are mine).

File # FTW02LA128: "I believe we had a *partial power loss* or anyway a *reduction of power* of some kind." When I taught PTS mandated techniques, one of my students who had just gotten his Private took off from our 6500' field with full fuel and a passenger in our flying club Piper Colt. He believed the engine had failed and made a very smooth forced landing. Then he ran into a ditch, curled the prop forward, and bent the crank. This accident was not his fault nor was it mine. We both carefully adhered to Practical Test Standards in his training program. I hadn't the heart to tell him that he had really made a forced landing with full power in a downdraft. I did change to the Langewiesche short field takeoff technique.

FTW02LA082: "The airplane's *engine then began losing power*, and the airplane began to sink." The investigation indicated no loss in power but he *believed* he had lost power.

FTW02LA128: "However, once airborne the airplane did not gain airspeed...sensed a *lack of adequate power* development." Perhaps he sensed a downdraft or downwind shear but had been taught to believe gauges over senses. I believe over integration of contact and instrument flying skills has led to the dismissal of many contact flying skills. When we takeoff in VMC we are in a contact environment. Failure to use and believe our five senses will lead to bad conclusions like the one above. He sensed a lack of zoom reserve in the form of airspeed. The sensations were low relative wind noise, laboring engine noise, the sloppy feeling of insufficient buoyancy, the feeling of too little relative wind resistance through the controls, the feel of the wheel near the stomach, the sight of too much blue in the windscreen, the smell of gas from the carburetor because of the high angle of attack allowing the relative wind to flow directly from the carburetor to his nose, and finally the sight of airspeed trending downward on his airspeed indicator. This airspeed indicator was the last clue to manifest itself and he would have done a better job of contact flying had he been looking at the ground rather than the lagging airspeed indications.

DEN02LA047: "I kept trying to force (the) airplane to fly, but didn't have enough airspeed." This says it all. You can't force a plane to fly. It will fly when it has sufficient buoyancy. That buoyancy is inadequately shown on an airspeed indicator. Zoom reserve in the form of airspeed means the airplane feels like it will not only fly but maneuver well. This feeling comes through experience but we at the Contact Flying and Aircraft actively teach our students how to recognize this feeling and what the aircraft wants to do.

MIA02LA105: "...pitching down to attain sufficient airspeed, but the aircraft *would not climb sufficiently*." We don't want to climb. We want to

CONTACT FLYING

fly level, in ground effect if possible, jump the trees or rudder turn around them, then fly level or down into a valley if one is available.

MIA00FA201: According to the student, "during the takeoff/initial climb, the aircraft got to an altitude of about 50 to 100 ft, and *seemed as if it did not have enough power to fly.*" No small trainer, especially a 152, seems to have enough power to fly. The designers expect us to have the training and common sense to use gravity, thermals, ground effect, and hydraulic lift when power will not carry the day. Natural conditions are infinite whereas engine power is finite.

LAX02FA120: "The airplane never attained more than 70 feet off the ground as it flew past the departure end of runway 1, and while still 'wallowing,' began a right crosswind turn." Seventy feet is not a lot more than five feet; and it takes away the natural help of ground effect.

ATL01FA082: "She stated the airplane looked like it was not able to get enough power."

ATL03LA003: "the pilot noticed a loss of engine power."

CHI01LA295: "airspeed 60 rotate up. Very shortly power began to decrease." This is very normal. The solution with low powered or overweight aircraft is to follow the rotate up with a push over hard to stay in ground effect.

All of these accidents involved behind the power curve flight where pulling back on the stick caused descent. These pilots did not understand that when we have lost our buoyancy or what is called zoom reserve in the form of airspeed, pulling back will cause descent due to excessive angle of attack. Until zoom reserve in the form of airspeed is re-established, the stick works backwards.

The common times we enter this reverse control situation is when we get too slow on takeoff (Vx may be too slow) and when we purposefully get behind the power curve on the apparent rate of closure approach. At these times of reduced buoyancy or loss of zoom reserve in the form of airspeed, pulling back on the stick will cause descent and pushing forward on the stick will cause assent. When the angle of attack is near the critical angle of attack as during the Vx takeoff or the apparent rate of closure approach, we will mush down prior to stalling. This mushing is useful on landing (Mr. Langewiesche calls this the stall down landing technique) but very harmful on takeoff. We avoid the mush on takeoff by always pushing over to stay in ground effect. When we think we don't need the extra zoom reserve in the form of airspeed on takeoff is probably the time we will need it.

In the chapter on Low Level Forced Landings I discuss a takeoff forced landing I had flying pipeline. I mention obtaining zoom reserve in ground effect every time I take off and that a possible forced landing is one reason for that habit, which I have not broken in thirty years. But this obsession results more from hundreds of takeoffs in very heavily loaded Pawnee and Callair ag planes. Whether bush or ag, we work up to heavy loads by increasing our loads gradually until we know the aircraft's limits in various terrain, density altitude, temperature, humidity, takeoff site condition, slope, weather, etc. Because our iterations are vast, we can reasonably work up to greater than POH gross weights. Thus we have vast experience practicing slow flight at very low altitude. And by insight or experience we soon realize the great heavy lift advantage of ground effect flight.

To get to a field while hauling 140 gallons over roads, dikes, fences, trees, electric service wires, high tension long distance transmission wires (we only go over if we can't find a way under) and trailer truck rigs on the highway, we are conditioned to abhor the thought of climbing, pitching up, or gaining altitude. Attempting a climb for no other purpose than to obtain altitude, such a normal thing for the low time or high altitude pilot, would be suicidal for the spray pilot with these loads. Without the help of thermal or hydraulic lift, which will be covered in the chapter on Low Power Mountain Flying, we cannot leave ground effect for more than seconds until some of the load is sprayed in the first spray swath or two.

Greg Simler, my manager at Underwood Aerial Patrol, once said, "Jim Dulin doesn't believe airplanes will climb." I am alive today because I don't believe, with any assurance, that airplanes will climb. It is just a survival mind set. Brain washing is effective and can be useful. These operations need not be attempted. We could haul 100 gallons and make a few more trips. We could purchase a bigger spray plane with a bigger engine. We could purchase more small airplanes and hire more pilots. Winston Churchill once said, "a man's life must be nailed to a cross either of thought or action." Airline pilots are men of thought. Old crop dusters were men of action. If you'll suffer another quote, this from the Kevin Cosner movie, <u>The Postman</u>, the old Vietnam veteran said he, "knew stuff." Men of action know stuff.

Many pilots believe that Vx results in a straight line incline. *See Figure 6.* This is not a C-152 making a short field takeoff over an obstacle. It is an airliner leaping a tall building in a single bound. While generally possible in powerful aircraft, this stabilized assent is not possible when power is low or load is high or conditions are powerfully influential. The downwind shear degradation on climb becomes more influential as the size of the engine decreases. Small, low powered aircraft do not climb like large, high powered aircraft and we err as educators if we try to make it so in the interest of standardized training.

Figure 6

In the AOPA Air Safety Foundation study of all instructional accidents (both dual and solo) from 1992 to 2001 they found that takeoff accounted for 12.6% of the dual accidents and 12.2% of the solo accidents. However dual takeoff instruction accounted for 14.9% of the total fatalities (#of accidents) where solo takeoffs accounted for only 6.7% of the total fatalities. Our instructors are doing their very best to teach a short field over an obstacle technique that is required to pass both the Private and Commercial Pilot Practical tests. It seems that meeting the Practical Test Standards can be dangerous. Could it be that students instinctively do not push the standards so hard when solo?

The introduction of V speeds came with the need to integrate contact and instrument training and promote the seamless transition to larger aircraft. But insistence on Vx for takeoff over an obstacle and 1.3 Vso for landing has degraded the art of contact flying using all of our senses. The Vx technique causes every low powered trainer to "appear as if it does not have enough power to fly." Using 1.3 Vso will cause every landing to go long including those into short fields and forced landings.

On takeoff with "zoom reserve" in the form of airspeed, our aircraft is buoyant and alive. We have maneuverability. We can turn left or right, go up or down, whatever. We are committed neither to mushing into the trees, should nature let go with a downwind shear, nor to flying straight into a brick wall with a dead engine. As Mr. Langewiesche continues in Stick and Rudder, on page 85, "The experienced pilot will therefore point his nose up only as high as absolutely necessary to clear obstructions. In fact, if the take-off is from a very tight field, he will often point the ship's nose actually at the obstructions---maybe at half the height of the trees. This does not help the airplane gain height, but it does help it gain speed. And the experienced pilot does not worry because he knows (without thinking about it) that speed and height are two forms of the same thing, and for various reasons he prefers to have that thing in the form of speed. And he knows that he can always at the last moment convert speed into height by pulling the nose up; the airplane will then 'zoom' and clear the obstructions."

Figure 7

With the crop duster you were watching, the objective was speed. *See Figure 7.* But please understand he was not looking at the airspeed indicator to determine it. He was looking at the ground (three feet away in the field, further away in the turn). But you correctly add, from the ground he can determine only groundspeed and not airspeed. Zoom reserve in the form of airspeed is obtained through a comparison of groundspeed, how far the stick is back toward our stomach (feel don't look), the pitch attitude and whether it is increasing or decreasing or staying the same, the angle of bank and whether it is increasing or decreasing or staying the same, the kinetic feeling of buoyancy and whether it is increasing or decreasing or staying the same, the kinetic feeling and speed of assent or descent or remaining level, relative wind noises and whether they are increasing or decreasing or staying the same, the sight of correct nose trim when the nose is moving at a horizontal speed proper for the angle of bank, the feel of correct nose trim when neither butt cheek is unequally loaded, the lack of relative wind rattle indicating proper trim, the willingness of a wing to come up smartly indicating proper rudder usage, the feeling of positive gravity more or less but always positive (no chemical coming up onto the windshield), and the lack of gas smell indicating too great an angle of attack (relative wind from carburetor to nose). Actually he is not looking for speed but for the positive feeling that he will go up smartly when he pulls back on the stick. After applying full throttle, the left hand moves to the spray handle. On pull up the spray handle is closed and the left hand moves to the dump handle.

By staying in ground effect until pull up the crop duster makes full use of Newton's Law of equal and opposite reaction. This extra push is the key both to takeoff over obstacles and to pull ups at the end of the field. This extra push is the key to short field takeoffs whether over an obstacle or not.

A Pawnee goes into the field at 100 MPH and pulls up at about 90 MPH. At three feet spray altitude ground effect is fully utilized. Toward the end of the field we use the bottom of the obstruction to judge the pull up point (A). *See Figure 7.* Then our eye shifts from the bottom of the obstruction to the top. The aircraft responds to the natural control input that follows. Over the obstruction the aircraft is leveled unless it still feels buoyant enough to continue up. Up (altitude) is a fragile commodity and we don't want to trade too much of our zoom reserve in the form of airspeed for this zoom reserve in the form of altitude. While still buoyant (lack of sinking sensation, good relative wind noises, no rattle, no gas smell, etc.) we level the aircraft and after clearing the wing we turn downwind (B). We spray crosswind and work from the downwind side of the field to the upwind side where possible. The angle downwind (B) depends on groundspeed or in other words the strength of the wind. Now we again pitch up (C), if buoyant enough, to loose airspeed and groundspeed trading airspeed for altitude. The slower we go the faster we will turn. As we begin to loose buoyancy in this climb (D), we turn sharply allowing the nose to fall through naturally trading zoom reserve in the form of altitude for zoom reserve in the form of airspeed (E).

This same crop duster turn, or what gun pilots call the return to target, was taught as the lazy eight in the old days. Now, unfortunately, the

CONTACT FLYING

inflexible PTS speed, altitude, and bank numbers make the lazy eight more of an instrument maneuver. The regulatory reduction in bank angle makes it awkward to manage the energy of gravity (turning sharply allowing the nose to fall through). There is simply no way we can pitch up to a designated pitch attitude and get the best altitude gain in all conditions. In an updraft the contact pilot naturally pitches up faster and to a greater pitch attitude. In a downdraft he pitches up less. The contact pilot banks sharply to get most of the turn completed before the natural pitch down of the nose increases the speed. He knows that turning with great speed causes load factors. This turn is a turn to target (crop row, enemy bunker, or runway) not a turn to heading. Turns to heading are instrument maneuvers and flying by reference to instruments limits energy management.

At the pitch up and roll in position (D) in the crop duster turn we are at our slowest airspeed (we have had the nose up) and slowest groundspeed (we have turned upwind) which allows us to easily pick up the next swath row (F) and to better miss obstacles going into the field. By storing zoom reserve energy in airspeed in the field and in altitude for the critical return to target, and by making this turn into the wind, the Ag pilot takes advantage of all natural energy. He uses ground effect to zoom over the obstruction (A), gravity to increase airspeed D-F) and headwind to reduce groundspeed. Attempting engine climb alone in a futile attempt to gain and maintain too much altitude, he would mush into the trees. Bad gusts affect him too, but he has maintained zoom reserve throughout by continuously trading speed for altitude and altitude for speed as energy appropriate. The only time airspeed was sacrificed was to zoom over the trees and to slow down in order to make the turn within sight of the field (A&C).

Understand that the altitudes and speeds are variable depending on conditions. While the turbine spray plane may get up as much as 800 feet, the loaded Pawnee never goes above 200 feet and often makes this entire return to target at less than 100 feet. But the operation is made possible by trading zoom reserve in the form of airspeed (in the field) for zoom reserve in the form of altitude over the tree for zoom reserve in the form of airspeed downwind for zoom reserve in the form of altitude in the downwind pull up for zoom reserve in the form of airspeed in the return to target.

Small aircraft cannot leap tall buildings in a single bound. We can, however, use ground effect in the field and over the runway and gravity in the energy management turn. At Contact Flying and Aircraft you will manage all available energy using contact flying techniques and fly little aircraft more effectively and safely.

On those troubling short field takeoffs over obstacles, we could have traded zoom reserve in the form of altitude (gravity) for zoom reserve in the form of airspeed but the training program mandated by the Practical Test Standards insists we chase an inefficient airspeed indicator rather than to use all visual, kinetic, aural, smell, and taste sensations. My stomach churns less using the ground effect takeoff. Have you ever seen a loaded Ag plane use Vx? We must train according to the Practical Test Standards, even if unsafe, to pass the Private Pilot Practical Test. Therefore we will find ourselves out of ground effect and sinking on many short field takeoffs. At this point we must get our eyes off the airspeed indicator, which is just now showing the sink (low airspeed) that happened more than a second ago, and use gravity to reestablish flying speed while trying to hit the tree. Then with zoom reserve in the form of airspeed we may zoom over the tree. Next we must level the aircraft above the tree to regain the speed lost in the jump. Finally we may have to descend into the valley beyond the tree if available. If no hole is available, we will just have to keep the aircraft level and hope we accelerate sufficiently to climb or maneuver away from further obstacles.

The pilot in the accident report MIA02LA105 who mentioned, "pitching down to attain sufficient airspeed, but the aircraft would not climb sufficiently," may not have tried to hit the tree. He may have pitched up less to try to attain sufficient airspeed as taught in takeoff and departure stalls. When he practiced takeoff and departure stalls he was probably at a high altitude (required by regulations). At three thousand feet the maneuver had no ground reference. It is hard for a pilot to push the stick forward when sinking with a real life ground reference. It is much easier to stay down in ground effect and pull up at the last possible moment to zoom over the tree. At Contact Flying and Aircraft we provide more training in ground effect, impart an understanding of zoom reserve, and teach what the aircraft wants to do.

Practical Test Standards that include more contact flying techniques would be helpful in preventing these inefficient energy management situations. We operate in an environment that is very unstable with aircraft that compare more to gliders than to heavy twins, yet the PTS requires our students to attempt stability, where in fact, flexibility may be required to survive. At Contact Flying and Aircraft we emphasize flexible reactions to natural effects, acceptance of the limitations of engines, and use of environmental aids to flight.

If we emphasize stalls so much that a low time student has flown half of his forty training hours with the tail low and the engine laboring, he might well believe that such an attitude is perfectly normal. He might even believe it is fine for the aircraft to act like that on takeoff or go around. He will realize that the aircraft *should not be flown that way*, and that it will lose zoom reserve in the form of airspeed and start sinking long before it stalls, only if he is taught to lower the nose *anytime and anywhere (except landing) he feels that sinking feeling, smells that gas, hears that engine labor, hears that weak and ineffective relative wind, and feels the stick or wheel near his gut.* Yet he may spend much of his training time climbing ineffectively and practicing stalls. He may not realize that in a real situation he could have hit the trees in a mush during that long period at a very slow speed waiting for the stall break that he is trained to react to.

The required short field takeoff procedure using Vx causes pilots to remain in any downdraft the longest possible time and therefore lose the most zoom reserve both in the form of altitude and in the form of airspeed. This may have happened in a Tulsa short field training takeoff FTW98LA294 where the, "student pilot established the airplane at the best angle of climb speed; however, the airplane's climb performance was degraded to the point that the airplane was only climbing at a rate of 100 feet per minute." I will cover natural conditions in greater detail in the chapter on mountain flying but we always want to spend less time in downdrafts and more time in updrafts.

We accomplish a net gain in zoom reserve, both in the form of airspeed and in the form of altitude, by flying fast in downdrafts and slow in updrafts. On takeoff it is critically important to fly through downdrafts quickly. If we accelerate in ground effect downdrafts have no effect on us except to push us horizontally. As the shaft of air descends, the ground forces it to flow outward in all directions. We may enter the upwind side of the outflow which causes an increase in airspeed but no descent (we are in ground effect). We may enter the downwind side of the outflow which causes a decrease in airspeed but no descent (we are in ground effect). Or we may be moved sideways.

Using energy management to maintain zoom reserve either in airspeed or altitude with the flexibility to swap airspeed for altitude and vice versa, gives us enough control of the situation that the outcome of the maneuver is not continuously in doubt. Had the instructor in the Cardinal at Cheyenne just allowed the nose to fall through naturally, in his turnaround to go east to lower terrain, he might have made a no load factor turn. He would have lost altitude by trading altitude for airspeed but he could have survived. While cautioning our students about maneuvering flight at low altitude, do we explain that low altitude maneuvering may be necessary for survival on any takeoff? In aircraft with limited power or heavy loads or both, stabilized takeoffs using V speeds do not tolerate much in the way of natural variables.

CONTACT FLYING

Contact Flying and Aircraft understands that the takeoff situation is fluid and may call for flexibility. Another short field technique is to stay in ground effect and miss the obstruction laterally or even go under wires. This is crop duster stuff, whether in the 235 horsepower Pawnee or the 1700 horsepower Air Tractor. It requires a flexible pilot and some training but can help anyone out of a sticky situation. That Air Tractor flies like a pig with 800 gallons of chemical and requires good contact flying skills. In a 12,000 pound Cobra (2500 over gross) we stay in ground effect and pedal turn around obstructions (We miss to the right if possible because we already have near full left anti-torque pedal to counter the tendency of the aircraft to rotate equal and opposite to the rotor. As we add collective pitch to the main rotor we have to add left pedal. When left pedal hits the stop and we continue adding power, we spin to the right. This can be managed by yawing right.

The same technique in airplanes is to rudder turn around the obstruction. We have to fight the banking tendency with opposite aileron, as allowing a wing to bank while in ground effect will drag a spray boom or cartwheel the aircraft. We lose a little airspeed because we are out of trim, but we are in ground effect and we are not killing speed by trying to climb.

Kids don't try this at home. Going under wires simply requires maintaining a six inch to two foot altitude, checking for traffic on the road, and not looking at the wire. We tend to put aircraft where we are looking. If the answer to the following Army Aviator Annual Written Test question is not obvious, don't try this maneuver:
In a high threat environment, which is the most effective way to deal with electric transmission line obstructions?
Fly over the wire at a pole.
Fly under the wire at a pole.
Fly over the wire between the poles.
Fly under the wire between the poles.

Flight Instructors and Examiners would do well to ignore me if training only instrument pilots in large aircraft. Our training system, however, pretends we all fly powerful airplanes in stable conditions. If we insist that Practical Test Standards are always possible we cloud our student's perception of what is really happening in light airplanes. If we overemphasize V speeds, stabilized procedures, shallow banks, following checklists no matter how unrealistic for small aircraft, control pressure rather than movement, level turns, limiting load factor by limiting bank rather than by allowing the nose naturally to fall through as the airplane was designed to do, and gaining safety through powerful engines and high altitude we prepare students well for instrument flight in large aircraft. At the same time we instill in them unrealistic expectations of the little airplane they are actually flying.

The Flight Instructor's Pocket Companion sets the proper attitude for transition to instrument flight in large airplanes but it might cloud the young flight instructor's perception of how training airplanes really fly. On page 159 of *Flight Instructor's Pocket Companion*, concerning takeoff over a 50 foot obstacle, we are taught, "At about 5 mph below take off speed, raise the elevator to obtain a low angle of attack. As soon as the airplane is positively airborne, adjust the attitude to obtain the recommended best angle of climb airspeed, Vx. Maintain that airspeed until the altitude exceeds the height of the obstacle. Then gradually lower the nose to allow airspeed to increase to best rate of climb airspeed, and then to the recommended climb airspeed. It is critical to not allow the airplane to descend in transitioning from one selected airspeed to another. Maintain a positive rate of climb throughout." While absolutely necessary to pass the Private Pilot Practical Test and while possible in good air with a good engine (if nothing natural happens), these instructions can cause students and instructors to believe incorrectly and execute incorrectly in real life takeoffs where nature intervenes.

According to Certified Flight Instructor Oral Exam Guide on page 4-9, "A takeoff and climb from a field where the takeoff area is short or the available takeoff area is restricted by obstructions, requires that the pilot operate the airplane at the limit of its takeoff performance capabilities. To depart such an area safely, the pilot must exercise positive and precise control of the airplane attitude and airspeed so that the takeoff and climb performance results in the shortest ground roll and the steepest angle of climb. In order to accomplish a maximum performance takeoff safely, the pilot must be well indoctrinated in the use and effectiveness of best-angle-of-climb speed and best-rate-of-climb speed for the specific make and model of airplane flown."

When Vx increases to Vy with altitude, it becomes nonexistent on the high density altitude takeoff. When 115% of gross weight, legal in Alaska, causes a higher angle of attack in level flight, Vx becomes nonexistent. At altitude or high gross weight we already have significant pitch up just to stay even. But, for a student unwilling to question his instructor, there is an implication here that a good engine and a good pilot will overcome any natural elements that come into play on a short field takeoff over an obstacle.

We can blame neither approved instructional materials nor our instructors for the misleading assumptions made in these texts. Their techniques for short field takeoff are absolutely necessary to pass the Private Pilot or Commercial Pilot flight test. The blame lies with a regimented and inflexible training curriculum which, in an effort to condition pilots for instrument flight in large aircraft, loses a common sense understanding of how little airplanes fly. The emphasis on power and precise application of a theoretical best angle of climb speed goes against the experience of all pilots. We all flew little airplanes first and we all know, without even thinking about it, that little aircraft fly quite differently than large, jet powered aircraft. Political correctness leads to inefficient resource management in government. In aviation it leads to excessive reliance on big engines and expensive panels where proper training could produce higher returns in safety. At Contact Flying and

Aircraft we recommend LSA tail wheel primary training using the more flexible and contact flying friendly LSA Pilot Practical Test Standards.

Conditions, not our engine, determine whether the airplane will indeed obtain the recommended best angle of climb airspeed but we don't want the best angle of climb at an early stage in the takeoff. We want zoom reserve, in the form of airspeed, before reaching the tree. By omitting anything about lowering the nose to remain in ground effect until Vy is established, these training manuals expect students to make some dangerous assumptions. First the student must assume that the aircraft will continue to climb, out of ground effect, to Vx. Second the student is asked to assume that the aircraft will, out of ground effect and in a near stall pitch attitude, climb above the height of the obstacle. Third our student is led to believe he can always accelerate to Vy without descending below the height of the obstacle. Finally the student is expected to believe that he can always prevent descent of the aircraft while "transitioning from one selected airspeed to another" and "maintain a positive rate of climb throughout." These are very dangerous assumptions to make in low powered aircraft.

What does the student think when we say it is critical not to allow the aircraft to descend? Does he think he should keep pulling back on the stick to prevent descent? He is at full throttle at this juncture. How do our students not allow the aircraft to descend if nature says, in the form of a downdraft, "DESCEND?" Should we teach our students to depend on their engine to prevent descent? Loss of zoom reserve in the form of airspeed means that pulling further back on the stick will result in mushing down or stalling even at full power. If we sense low zoom, we must push forward on stick and or throttle. However, the throttle is full forward. When airspeed digresses to this point our student must push down on the stick to go up (at low speed the elevator does not elevate, but just the opposite). Our contact students understand that they induce drag by pulling back on the stick and that the aircraft would not do this, but would do just the opposite. Our students are taught to simply let go, when unable to create more lift by pulling back, and thereby allow the airplane to do what it wants to do. When behind the power curve it wants to pitch down and use gravity to perform better.

Our contact students understand that our objectives are quite different in the takeoff phase than in the landing phase. In chapter four we will learn how to manage induced drag both with flaps and with pitch attitude on landings. But on landing we are not interested in developing zoom reserve either in altitude or airspeed. Rather we wish to dissipate both completely by the touchdown point which should be the very beginning of the landing area. Our objective on takeoff, however, is to first gain zoom reserve in the form of airspeed and then in the form of altitude.

A word here is appropriate to young instructors. Ladies and gentlemen I know you are not getting textbook results from your student's short field takeoff efforts. I know you are not getting results because I have extensive experience with natural elements, I've flown a long time, and I've seen you out there having poor results. It is not your fault nor is it your students fault. This particular maneuver does not work unless conditions are prefect. I know you are constrained by the PTS and have to teach Vx even though you understand its limitations. At least teach your students to level the aircraft when things get dicey. The 50' obstruction is not really there. Teach your students to level the aircraft when necessary and live. If the obstruction is really there and you are in a downwind shear (sinking), level the aircraft to fly out of the shear as quickly as possible. When you are sinking and you shouldn't be, it is not likely an engine problem. You are in sinking air due to a downdraft off an obstruction or a temporary downwind (or less upwind) shear. Whatever the cause, you are going down. Slowing up further is not the thing to do here (i.e. pitch up to Vx). Flying level or downhill is the thing to do here.

At Contact Flying and Aircraft we find that our experiences, not stabilized V speeds, keep us alive long enough to put the POH on small aircraft into proper perspective. In the early years of aviation there was neither POH nor V speeds for any small aircraft. We had fewer takeoff and go around fatalities at that time. V speeds are cool big airplane stuff but they are a dangerous way to learn to fly small aircraft. With experience we learn that our airspeed indicator works very poorly below cruise and is a major distraction in ground reference work (which includes takeoff and landing). At Contact Flying and Aircraft we cover up the airspeed indicator from first flight to solo because it diverts the student's attention from the business of flying the aircraft based on how it feels, smells, sounds, the apparent rate of closure with the ground, the stick position, etc.

Theoretically zoom reserve can be stored in the form of altitude just as efficiently as airspeed. But man, no matter how hard he tries, is not a machine. When the time comes to pitch down and try to hit the tree he may delay or freeze. Also there is an instrument delay in airspeed which will delay his reaction if he depends on this trend instrument. He will certainly wish he had stayed in ground effect long enough to build zoom reserve in the form of airspeed. If we simply stay in ground effect until zoom reserve in the form of airspeed is established, we will be ready if nature throws the negative shear. In ground effect a downdraft cannot take us down. Instead it will throw us sideways. This happens because air is a fluid that flows toward the area of least pressure. The downdraft pressure outflow will be horizontal or back up creating an updraft. We experience this horizontal turbulence often in the field while spraying crops. Also, we will spend less time in a takeoff down draft at a faster airspeed.

The pitch attitude and V speed technique taught student pilots in little Cessnas is entirely appropriate for the airline pilot who scares me with the violent pitch up to sickening pitch attitudes. Here zoom reserve is maintained almost entirely by gas production (rocket power). This is a human engineered and prearranged use of natural laws. When the fire is burning sufficient zoom reserve is available to pitch to Vx. When the fire goes out the jetliner comes down. This rocket force is sufficient to

CONTACT FLYING

overcome most but not all shear problems. Similar pitch attitude and V speed techniques required by the PTS are not appropriate for small, low powered aircraft. We can not "leap tall buildings in a single bound."

Few become airline pilots. Because they are responsible for so many lives, their small number does not detract from the need to emphasize instrument integration and transition to larger aircraft in their primary instruction. In fact it seems counterproductive to start them in small aircraft at all. I am not suggesting that the training program change for these chosen few. I simply ask that general aviation pilots and the government people who say they support them recognize and value the art aspect (using all of our senses to determine our ships relationship to the earth rather than just instruments) of contact flying. I seek respect for those who view the small aircraft as a fine tool rather than a weak beginning to bigger and more powerful aircraft. If the entire training system from pre-solo through Commercial Flight Test was based on small aircraft in the contact environment, all but airline training programs could go back to teaching people the art of flying. Those aspiring to large aircraft corporate work or the airlines could undergo an advanced training program similar to today's primary training program except that large, powerful aircraft would be used from the beginning..

Surface conditions, terrain, obstructions, density altitude, slope, and wind conditions are some of the many factors that lead to takeoff accidents resulting in fatalities in *www.ntsb.gov*, but one factor is almost always present: a pilot who was blinded by the misguided belief that airplanes *always* climb more in a set distance at Vx and always climb faster over time at Vy. Small, low powered aircraft do not always climb at Vx or Vy. Sometimes surface conditions, obstructions, density altitude, lee side downdrafts, thermal downdrafts, compression shear on the windward side of mesas, slope or wind conditions more than overcome the total power available from the engine.

This does not mean we cannot fly small, low powered aircraft in mountain or bush operations. It just means we must be more aware of natural conditions than of engine performance. Using ground effect can increase takeoff performance 20% on a poor surface and 10% on dry, level, hardstand. Simply going downhill during or just after takeoff can increase performance 100%. Using ground effect to build zoom reserve in the form of airspeed will deal with obstructions many times better than Vx. Leaning before takeoff, taking off toward lower terrain, and staying in ground effect until reaching lower terrain will more reliably deal with density altitude than installing a larger engine. Taking off down slope rather than up slope simply means flight is possible. Staying in ground effect until zoom reserve in the form of airspeed is achieved, will effectively deal with negative wind shear and downdrafts where Vx just sets the pilot up for disaster.

Flight training is a study of belief, not just a study of fact. If we will look about us and recognize the true power available to us, the Almighty will help us deal with our inadequacies (engine or otherwise). Powerful engines and small wings teach power management, but small engines and large wings teach flying. Systems management teaches instrument flying, but natural energy management teaches contact flying.

Let's look at takeoff training in light tail wheel aircraft. I preflight with my students the first day only. Thereafter they do their own. They are not encouraged to believe this ritual prevents anything from going wrong with the aircraft. They are encouraged to get to know and support their mechanic. I demonstrate once and then talk. My students learn by doing.

While the before takeoff checklist is important there is one thing on it that irritates me. The takeoff trim indicator position, if utilized, will put the aircraft in a normal climb. This makes the normal climb attitude the baseline control feel for beginning students. Normal climb attitude should not be the baseline control feel for beginning students. They assume this is a normal attitude and control feel. In the interest of safety

CONTACT FLYING

we need to teach cruise flight as the normal attitude and control feel. We should have set the trim at cruise on the last flight and to takeoff safely we must leave it there. We should mark normal cruise on the trim indicator and ignore the takeoff setting. In the interest of safety, our students need to believe that nose level is the normal attitude and that cruise flight is the normal control feel, normal relative wind sound, normal buoyancy, and normal smell for small aircraft. I understand the need to teach trim usage for instrument work and transition to larger aircraft. I believe however, that this training should follow complete indoctrination in the basic feel of a small aircraft trimmed to neutral in contact flight.

I never allow the contact student to change the trim except to adjust for normal cruise. In contact flying we spend very little time on long climb outs and long descents. The amount of time we will have to hold backpressure on the stick is minimal. Cessnas are set up so that the trim setting for full flaps is the same as the trim setting for cruise. Is somebody trying to tell us something here? Yes we need to use full flaps on all contact landings and no we don't need to mess with the trim to land. If we mess with the trim on landing we mess up our own landing and we mess up the trim for the next guy on takeoff. In contact flying with small aircraft, we leave the trim alone except to trim the aircraft for normal cruise flight.

If we will just leave the trim alone except to trim the aircraft in cruise, we will have a control position and feel we can depend upon in an emergency. Thus our students will have a control feel that teaches kinetically and will not be a "false teacher" to them. On takeoff with the trim set for normal climb, the control feel is like a wet noodle and does not give the student a good feel for the amount of buoyancy and the amount of zoom reserve not available to him in the form of airspeed. The position of the stick or wheel is wrong and to make it right he has to push forward. Pushing forward on the stick to keep safe zoom reserve in the form of airspeed is awkward. When trimmed for climb he will need to push the nose down to stay in ground effect. Thus to build zoom reserve in the form of airspeed he will have to fight against the excessive up trim.

The combination of checklist nose up trim for takeoff and possible autopilot activation (ATL02LA166) caused a student pilot to close the throttle and make an emergency off field landing because, "the nose was up and the airplane was coming down."

Tail wheel airplanes were designed with poor mechanical brakes and little tail wheels for good reason. Good Cleveland brakes are hard on props and wings and teach reliance on good Cleveland brakes. There is a simple answer to the brake problem in small tail wheel aircraft: don't use them. Use rudder (all of it) and power (a lot then none) to taxi. Directional control on or near the ground is all rudder control, no stick. It's dynamic not static. An athlete shifts his balance from the ball of one foot to the ball of the other foot continuously. We don't learn to balance a broom on our hand by holding our hand static. We move our hand dynamically, very much at first then less as we get control. Continuously pushing alternating rudder to the stop will result in a straighter taxi than moving the rudders only a little. Also the student needs to know where the limit (stop) is. Crisis time, when he is just about to lose it, is not a good time to be searching for the limit. Small tail wheels (or skids) help keep the tail behind.

We keep the stick back to the stop when taxiing because we have no rudder control until we gain enough speed to push the tail up to level. We must rely on tail wheel steering. When we gain sufficient speed to get the tail up we push it up smartly to avoid a tail low (no tail wheel and very little rudder effect) condition. We taxi slowly, especially on pavement, because with poor brakes we might have to shut the engine off to keep from hitting something with the prop. Installing good Cleveland brakes, on the other hand, may put us on our back in an emergency stop. Grass makes taxi and takeoff a snap because the small tail wheel mires in a bit helping keep the tail behind us where it belongs. On pavement, however, a strong gust can send us totally out of control requiring an engine shutdown.

We keep the nose on the yellow taxi line by alternating full left and full right rudder rapidly. Moving the nose left or right requires only slight alterations in this rhythm. If in side by side seating, the nose is straight out between our legs. If the taxi line appears to be between our legs our tail wheel (or nose wheel) is on the line. Step out to see this sometime. By the way, the line appears to be between the instructor's legs as well. If we look over the prop we will line the aircraft up crooked because of the angle between our eyes and the prop.

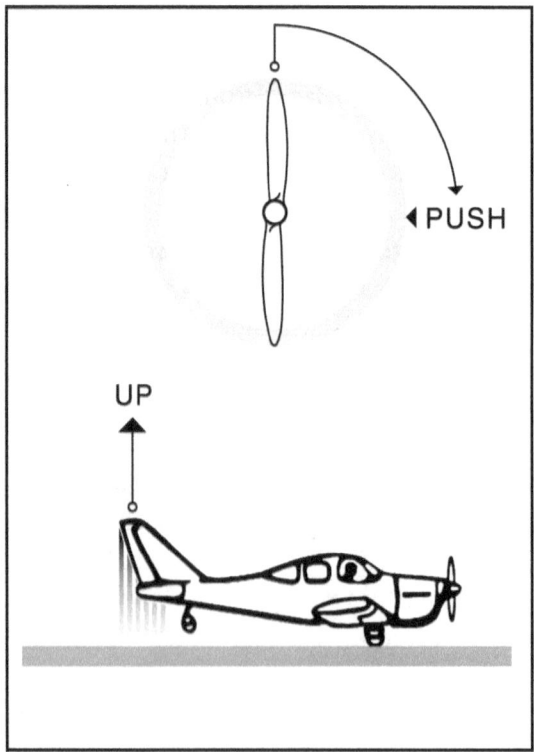

Figure 8

Staying ahead of the airplane, while gaining enough speed to get the tail up, requires lots of rudder work and knowledge of the effects of precession. *See Figure 8.* Unlike tricycle airplanes, which sit on their gear in the proper takeoff attitude, tail wheel airplanes cannot accelerate quickly in their tail down high angle of attack. Therefore we must push the tail up level with forward stick as soon as we have sufficient relative wind. It happens quickly in a strong headwind but slowly in calm air or a downwind. We walk the rudder a lot then less as we speed up and the rudder becomes effective. When we gain enough speed we push the tail up smartly. We don't wait until the tail comes up on its own. Waiting would give us a longer tail low transition time. Most ground loops happen during this transition. During transition we have insufficient rudder control yet our tail wheel is off the ground. Large turbine Ag aircraft eliminate most precession by using a very long tailwheel structure. This makes the aircraft sit almost level like a tricycle aircraft.

Rapidly pushing the tail up on little airplanes, however, causes the nose to jump left because this rapid leveling of the aircraft is like pushing on the top of the prop's tip path plane. This force takes place ninety degrees away in the direction of prop rotation (see illustration). Pushing the tail up is really like pushing on the right side of the prop's tip path plane. The nose is pushed left. This must be anticipated and corrected with dynamic right (right, left, right, ... but favoring right) rudder application as we push the stick forward smartly.

What I mean by dynamic right rudder application is that while alternating right and left rudder pressure vigorously we simply favor the right a bit. When we instructors talk about pushing right rudder or pushing left rudder we are thinking statically even if we fly dynamically. To keep the nose straight, beginning students must walk the rudders quite a bit to maintain a dynamic balance. We err when we scold them about over controlling as if we never had to do it that way. It's just like riding a bike. We wobble a lot at first then less as we speed up. This dynamic control does not go away, we just hide it a bit as we mature. The professional

CONTACT FLYING

tennis player doesn't even try to hide it. Watch him dance back and forth on the balls of his feet as he waits for the service. Should he find the prefect spot and sit on it, the service will go by him before he can react even though he is in prefect position. If our dance goes completely away we ground loop. Some military instructors insist on maintenance of the "sweet spot" or "hover button" that will hold a ship perfectly stable. They know it doesn't exist and they confuse their students by demanding it.

Using the controls is like finding the target with an area type weapon. It requires dynamic adjustment of multiple rounds. Since there are no accurate sighting systems for these weapons to achieve first round accuracy and no smart computer guided shells or rockets, adjustments must be made based on impact of early rounds. The experienced gunner may even adjust based on observation of the trajectory of early rounds leaving the weapon on the aircraft. We call this adjustment based on the observation of the flight and impact of rounds bracketing or "burst on target."

Flying is similar but requires constant corrections. Keeping the nose straight requires dynamic application of variable rudder pressures. The exactly right place to position the controls simply does not exist in aviation. It is a dynamically fluid art. Thus we have to walk the rudders when we taxi, when we start the takeoff roll, when we push the tail up, when we lift off, when we fly, etc. If we move anything, we move the rudders. If we don't move anything, we move the rudders. In the movie, <u>Butch Cassidy and the Sundance Kid</u>, Sundance is asked by the old miner to shoot his pistol statically. He holds the pistol static and uses the sights but cannot hit the bottle. Conversely, when he is allowed to move dynamically, he hits the can with each shot. "I'm better when I move," is his classic line. What he really means is, "this pistol is too short for an illogical scatter brain like me to sight and shoot. However through trial and error I have developed the instinct to just know where to point the pistol. Constriction of my movements or any logical sequential attempt to execute them will simply mess me up." Therefore, if we want to fly tail draggers and not ground loop we need only remember, "We're better when we move."

Now that we have the tail up acceleration is rapid. We keep walking the rudders but, as we get more relative wind across the controls, we notice that less movement is required to have great effect on the nose. We also notice that while less movement has more effect, more pressure is required to get this less movement. We wiggle the stick fore and aft to keep the aircraft level and notice that here too less movement and more pressure is required to keep things right. *See Figure 9.*

Figure 9

When we feel buoyant enough to lift off and begin to pressure the stick back a little, the nose goes left again even before liftoff. As we pitch the nose up, even before the mains leave the ground, the relative wind comes at us from below. This pitch up causes the right swings of the prop blade to fly at a more efficient angle of attack than does the opposing blade on the left side of the nose. Again we must anticipate this with dynamic right rudder. Anytime things aren't going right, then we move something just to see what it will do. If we don't like what it does, we put it back. We're better when we move. We simply cannot learn the controls unless we move them.

While we are not teaching and certainly not emphasizing numbers, we need to give the student a general idea of how things progress in takeoff and landing and when to be concerned with progress. If conditions or engine problems cause the aircraft not to be able to get the tail up by the 1000' point in the takeoff roll, we should abort. Small aircraft are weakly

CONTACT FLYING

powered but light. They can get off in ground effect quickly but tend not to climb well until sufficient zoom reserve in the form of airspeed is achieved. We need to get off into ground effect quickly or quit! We need not worry about getting up to some arbitrary altitude quickly as the aircraft will perform considerably better in ground effect.

OK! We've got the thing off the ground quickly but it sounds weak, it smells like gas (relative wind from below is moving fumes up to our nose level), it feels mushy and wobbly, and it is not going up rapidly. It is weak. That's a 65 horsepower engine up front. Lowering the nose to stay in ground effect makes things sound, smell, feel, and look much better. In more powerful aircraft like the Cessna 180 we have to push the stick forward fairly hard or we'll fly right out of ground effect. While dynamically walking the rudder to bracket the nose laterally (no ailerons) we also must alternately apply fore and aft stick pressure to bracket the nose vertically.

No coordinated aileron is needed on takeoff. Any aileron use is simply to keep the wings level. By wagging the stick fore and aft we maintain a steady three to five feet altitude. By walking the rudders we maintain directional control. When we feel we have sufficient zoom reserve in the form of airspeed (this comes only through experience), we shift our eyes from the bottom of the first obstruction at the end of the runway to the top of it. The hand follows the eyes applying backpressure to zoom over the trees with the outcome of the maneuver never in doubt.

I know someone in the back is going to ask that question; they always do. "What about rotation speed? Climb speed? Vx, Vy, etc.? Just give me some numbers. I'm a smart student and I can study these numbers at home." Like I tell my crop dusting students, there is no magic number in flying, no sweet spot. It's just not going to happen in a dynamic world. I could give you some numbers but I'd be lying. Natural elements are bigger than little engines. I cannot guarantee results based on V speeds. Conditions are too variable and "I'm sorry captain; we just don't have the power." If the gust spread is 20 and you climb at 60, nature has temporarily increased or reduced your performance 33%. For a rocket propelled airliner climbing at 200 only a 10% increase or reduction is experienced. You mean you're trying to get a dynamic handle on this narrow geared little demon and you want to look at a 3" diameter gauge inside the airplane? I'll just cover that thing up so it doesn't distract you.

The safest way to be sure we will have zoom reserve in the form of airspeed, at the critical juncture where we will need it (at the tree), is simply to stay down in ground effect until 1/8 mile short of the obstruction. While it will save tin and skin, you may encounter regulatory problems concerning this technique. I'm an instructor not a lawyer. If you hope to fly airlines or charter you need to talk to a lawyer concerning this technique. His advice may kill you some day but at least you will be politically correct using the Vx approach to short field takeoffs.

When we are high enough to clear the wing (turn without hitting something with the down wing) we make a level or descending turn (depending on speed) toward our destination. Climbing turns near the ground are dangerous unless we have sufficient zoom reserve stored in the form of airspeed. If we are fully buoyant (at cruise speed), we make an energy management turn. We lift the nose, with wings level, and then turn the nose onto the target. We allow the nose to fall through naturally, manage the tuck, level the wings, and pull up out of the resulting dive.

I have observed very difficult helicopter takeoffs in tight LZs in Vietnam. Our slick drivers would allow too many Blues (members of the Blue Platoon in every Air Cavalry Troop) to pile onto the Huey when they were in trouble. With the tall trees only feet in front of them (hover hole) they simply had to pull collective pitch until further power application would cause rotor bleed. Slick drivers fear loosing turns more than enemy bullets. Rather than apply left pedal to counteract the torque resulting from pulling max pitch/power, they simply allowed the aircraft to turn right in a spiral up out of the hover hole. Just applying that left anti-

CONTACT FLYING

torque pedal to keep the aircraft straight would have caused loss of turns (rotor deterioration) and a mush into the trees.

A similar technique is possible in fixed winged aircraft if we can stay in ground effect. We can rudder turn to miss the obstruction if we cross control the aileron to prevent the wing striking the ground. We will lose energy in the slip, but we more than make up for it by staying in ground effect.

Kids don't try this at home. Old crop dusters know they can get around an obstruction in the field using this rudder pedal flat turn. They also know that the distance from the leading edge of the wing to the prop spinner is less than the length of either wing. We can and have circumnavigated obstructions horizontally using this pedal turn and returned to the same crop row beyond the obstruction. Just before the right wing hits a tree in the field we push much left rudder and enough right aileron to keep the wings level. As the aircraft yaws left we push much right rudder and left aileron to stay near the tree. We will not hit the tree because the nose, now pointed at the tree, is shorter than the wing. Once we have cleared the tree we push much left rudder and right aileron to continue in the same crop row beyond the tree. Even the kids flying very long nosed turbine Ag airplanes have never seen this.

Chapter 3

DUTCH ROLLS

In the early 80s I was the FBO (Fixed Base Operator) at Buffalo, Missouri. I had a small flight school and charter operation but my bread and butter was my crop dusting operation. Midway through my second crop dusting season I crashed my Pawnee. As with all crashes it resulted from a combination of errors on my part. I was working out of a sixty acre pasture with four inch grass. I had just loaded one hundred twenty gallons of Parathion in water into the hopper when a couple more farmers showed up (two were already there).

It had been a very heavy and profitable alfalfa weevil season but everyone was backed up. Southern Missouri stock farmers (called ranchers further west) don't spray until the second stage of the infestation when the second brood is entirely hatched. At this stage they only have to spray once getting the large worms of the first brood with the small worms of the second brood with a heavy dose of Parathion. The problem with this single spraying, rather than spraying both broods, is that in bad years the fields can be damaged if not sprayed very quickly when the second brood hatches. Of course in a bad year everyone needs spraying so we get behind.

I was angling across the 60 to get slightly more than 3/8th mile for takeoff. As the two new farmers pulled their pickups behind the other two, it began to rain lightly. I should have dumped twenty gallons from the hopper because of the rain. I decided I would try it and dump on takeoff if necessary. The tail came up slow and she came off groaning but I felt I could stay in ground effect and jump the fence. I got above the woven wire and should have cut the barbed wire but by angling corner to corner to gain ground I also hit the top two barbed wires at an angle with my left gear first. This swung my right wing and gear around and pointed me

CONTACT FLYING

straight into a pile of stumps. I destroyed the left gear and left wing and messed up an already hectic day. I needed another airplane to finish the season.

My mechanic knew a crop duster in Mississippi who needed to buy a Stearman in Wisconsin for the engine. I was sent up north of Eau Claire to buy it if as advertised. I would be able to use it to finish the season and then deliver it to Mississippi.

The Stearman was duck taped and bailing wired together but it had a beautifully built up Aero 450 hp Pratt and Whitney engine, and the Mississippi crop duster said to buy it if the engine was good. I bought it. Some fabric under the upper wings was still flapping in the wind and the tail wheel was controlled by bailing wire connected to the rudder horns but the Aero engine buildup was a thing of beauty. I got in and headed for warmer weather. At the first airport I stopped and hitched a ride to town to buy insulated coveralls. It was a cold trip back to Buffalo with many stops for coffee and to warm up.

Most Army Air Corps Stearman Kadets had been converted to crop dusters. The front (student) seat was removed, the structural cross tubing behind the front seat was wrapped with fiberglass, and a two hundred gallon fiberglass hopper was fabricated in the void around that structural cross tube. A big problem with the Stearman was the four inch steel lip mounted on top of the hopper to increase volume and hold the lid. With my helmet touching the top wing, I could not cushion myself up high enough to see straight forward in that airplane. I asked the old timer in Mississippi, who was allowing me to use his Stearman, about my inability to see the field I was diving into. He said casually, "well you've got to look out around the hopper as you turn into the field."

Another problem we crop dusters had with the Stearman hopefully does not affect the extremely clean show planes out there today. On takeoff you had to close your eyes as you pushed the tail up on takeoff. This was because the relative wind necessary to get the tail up also created enough venturi effect on the open cockpit to suck the dust up from the floor and past your face. As soon as you felt dust in your face you opened your eyes again. It was a timing thing.

The real joy of the Stearman was that big, smooth Pratt and Whitney engine up front. After getting over being scarred by the fire flying past my right ear (you can see this at dawn or dusk) I learned to love the engine. This was the only crop duster I ever flew that would climb comfortably with a full hopper in mid summer. Also it was the easiest engine to start I have ever experienced. I would crawl up on the wing to turn the big mag switch to both and set the throttle. Then I would crawl down and walk around to the huge Hamilton Standard prop. I would give the prop a good swing, walk around and crawl into the cockpit, strap in, and put my helmet on while the big nine cylinder engine puffed and sputtered to a start 100% of the time. Radials are timed for full power so they run a little out of time at low RPM. The long, heavy prop has enough inertia to keep it going around until the engine begins running on most of the cylinders.

Nearly every Stearman that survived its post WWII crop dusting career with an airworthiness certificate and a few pieces of tubing intact is now a beautifully restored two-seater again. They are marvelous to look at but impossible to fly without good rudder usage.

I could put this airplane in a ninety degree bank to the right but the nose would not move one inch right until the right rudder was pushed. The Stearman I sprayed with did not have ailerons on both sets of wings. We pull up to get out of the field and slow down to turn. The slower you go the faster you turn (but you have to bank aggressively). Getting the wing back up after the steep bank required standing on the rudder. This is true, to a lesser extent, of any older airplane because they have no frieze aileron, bungee, or wing engineering to overcome some of the adverse yaw.

CONTACT FLYING

If you trim your turns by reference to the ball, you are flying by reference to instruments. We simply cannot learn how to maneuver the aircraft in low level contact flight unless we learn how to move the controls smoothly but also vigorously and dynamically. We cannot move the controls vigorously and dynamically by reference to instruments. Contact flying students need to learn from the first lesson that vigorous, dynamic movement of the controls is possible and achievable in maneuvering flight.

If you want to know how to fly, remember how you learned to ride a bike. Did you sit ridged and upright on the seat, hardly moving your body except to pedal in a "stabilized" manner? No you wobbled back and forth dynamically because that was necessary to balance yourself. Here's that theme again; we're better when we move.

Every time you take out a different aircraft (or the same old one) do a few Dutch rolls. If your airplane is modern, don't try to keep the bank minimal. Bank 45 degrees left and right while holding a distant point between your legs. That's exactly how you learned to ride a bike and it works just as well in learning or relearning how to fly a plane. Lean on the rudder as you move the stick or wheel.

After demonstrating Dutch rolls once, I turn the airplane over to my new student on his first exercise of his first lesson. After he can make forty five degree banked rolls left and right (less bank is needed in older aircraft) while holding a distant point between his legs, I talk him through turns, climbs, descents, rectangle course, pattern and landing. This is in the normal one hour first lesson. He does not need to see me practice a lot of turns. He knows how to do Dutch rolls so he knows how to turn with a little verbal help. If he knows how to turn he knows how to fly straight and level so talking is all that is needed to teach him to climb, descend, and fly a pattern.

Low time students need very little demonstration time. They can make much better use of the time for practice. Nor will acquiring so many hours watching an instructor fly give any student a good feeling about his own ability to fly. Flying is one of those learn by doing things. We instructors need to let them just do it. No golf pro could stay in business spending more than one percent of his student's time hitting balls (demonstrating).

Older aircraft are better trainers because they have no wing warping, frieze ailerons, or other means to make them turn with aileron only in a shallow bank. But I'm not saying the Ercoupe design is wrong. Mr. Weike got it right. The Ercoupe will make shallow and steep turns without slipping or skidding using ailerons only. Modern aircraft will do this only at a shallow bank and always out of trim. This is confusing to the student. Why, says he, can I usually make good aileron only turns yet get into real trouble sometimes using this technique? Older aircraft will not confuse him at either fifteen degree banks or forty five degree banks because, at any angle of bank, without rudder there will be no turn. The older aircraft has no wing warp, bungee connection between ailerons and rudder, frieze ailerons, or any other system allowing medium bank turns using ailerons only. If we must train students in confusing semi-modern aircraft, Dutch rolls must be practiced at forty five degree banks. They are pointless at any bank less than the bank necessary to overcome the protection of the wing warp, frieze aileron, or bungee system. That is why Contact Flying and Aircraft uses older tail wheel aircraft for zero timers and Ercoupes for old guys with experience but no medical.

Dutch rolls teach rudder usage. Modern aircraft do not insist on good rudder usage until the outcome of the maneuver is in doubt. Modern aircraft have some system to overcome some adverse yaw with aileron only. This can be warping of the aileron portion of the wing to fly at less angle of attack. Another method is to hinge the aileron so that parasite drag is produced on the up aileron to balance the induced drag of the down aileron. Many Piper aircraft have the aileron and rudder connected by a bungee cord that can be overcome in order to slip the aircraft. If we do not have the necessary rudder practice we will be unable to make sixty degree banked turns near the ground to miss obstructions or to turn around in narrow but deep mountain valleys.

The basis of all maneuvering is smooth horizontal nose movement (rudder control) and gravity energy management (pitch control). Without rudder control equal to the intensity of the maneuver, adverse yaw will not be countered. Unless we pitch up (from cruise) and then allow the nose to fall through in the turn, we waste energy and create load factor problems.

In an emergency we will fly as we have flown. This is just human nature. We will play the same game we have practiced. We will not rise to the occasion in some mystical way. If we have been landing long because the runway at our airport is more than sufficient, we will land long on any forced landing we make regardless of the circumstances. If we have not been using the rudders in shallow turns we will not use them in a steep turn to miss an obstruction and therefore we will hit it unless the Almighty intervenes. At Contact Flying and Aircraft we will make all landings on the numbers using the apparent rate of closure approach. Our students use older aircraft that require full rudder usage.

We've all done Dutch rolls sometime but we just don't practice much. When we spray all summer we get plenty of practice but midwinter may find us rusty. On a long cross country or even while flying pipeline it is easy to pick a distant target, put it between our legs to line up straight, and alternate forty five degree banks left and right a few miles. We simply start a turn. Once banked more than forty five degrees we reverse the turn through level to the same bank in the other direction while holding the target between our legs. When we fail to use sufficient rudder (especially when bringing a wing up) adverse yaw will drag the nose away from the bank. We have to correct by pushing the nose around with rudder. We don't let the nose push us around; we push it around. If the nose hangs, push it around. It's just like pushing the pedals on a bike or skiing. We lean left and right even when we want to go straight. We lean left and right on stick and rudder to continuously push the nose straight down the fall line toward the target.

If we fly modern aircraft, we will find it necessary to lead rudder. This is only because we don't think to use it. We haven't had to. By increasing the bank we will force ourselves to use the rudder. We especially have to lead rudder out of the turn to get the wing back up. Mr. Langewieche talked about the down wing aileron "going out" when insufficient rudder was used to bring it up to level. Perhaps this doesn't happen on modern aircraft but it still happens on older aircraft and it is a dangerous thing at low altitude to have a wing drag back instead of coming up smartly.

When the nose comes off the target in the direction of the bank, we are pushing the nose around too much. This is good initially to counter hours of nonuse. Usually, however, the nose hangs up and falls off away from the bank. In this more common case we are not using sufficient rudder. Again this is most common when trying to bring the wing up. We must practice getting the wing up smartly because failure to do so leads to problems when maneuvering vigorously near the ground.

Once the student has mastered the Dutch roll he knows how to turn the aircraft. He starts as if he were going to do a Dutch roll. After rolling the aircraft into a 30 degree bank, he releases the controls. He is now in a coordinated turn with his hands and feet removed from the controls. The nose will fall through to regain cruise speed because some of the lift is pulling the aircraft around.   To prevent this dive he will apply backpressure to the stick. This will happen naturally, not requiring demonstration, at 500' AGL. It will simply occur to him, perhaps without him even thinking about it, not to let the aircraft descend into the trees. Flying is not a mystery. We must dispel any mystique by relating flying tasks to tasks the student is familiar with. It will come naturally to our students when we let it.

CONTACT FLYING

At this juncture two other lessons about turns need to be taught: holding altitude is not necessary and the slower we go (airspeed) the faster we turn. Thus the energy management turn involves pitching up slightly to slow from cruise, initiating the turn, allowing the nose to fall through naturally, holding light backpressure to prevent the tuck, leveling the wings to roll out on the target (the ground target and not a number on a gauge), and pulling up wings level from the resulting dive. High altitude level compass turns are necessarily taught at an early stage of instruction but we should not give the impression that this is the only way to turn. If our student moves on to agricultural, pipeline, bush, light sport or even recreational flying any dogmatic need for a level turn and or compass heading will rise up to bite him.

Some years ago a student related to me that I was the first instructor to tell him how to use the rudders in a turn. Theoretically we need no rudder or aileron when in a coordinated turn. While theoretically we don't need rudder and aileron in a coordinated turn, we almost always end up needing them. Whether tense from the newness of flight or we just get sloppy, we need them. Nor will we usually apply rudder when a gust causes one wing to speed up or slow down. We tend to move the ailerons a lot and the rudder a little if at all. In rough air we must try to leave the ailerons alone and bring the wing back up with rudder only. Sometimes we set crooked in the seat causing uneven rudder pressure. For safety of flight and comfort, rudder must be applied anytime the aileron is moved or anytime the aircraft goes out of trim. Think of rudder as the trim control because that is what it was designed to do. Conversely, if things get so out of trim that you hear prop cavitations, relax on all the controls to allow the ship to re-trim herself.

Flying is dynamic. It's monkey see, monkey do. When we see the nose speed around the turn too fast (a skid), we slow it down with opposite rudder. When we see the nose hang up, failing to follow the bank (a slip), we must push the nose around with rudder. Flying is dynamic. It's monkey feel, monkey do. If we feel more weight on our right butt cheek than on our left butt cheek, we push right rudder. If we feel more weight on our left butt cheek we push left rudder. We push on the leg attached to the butt cheek that is being rammed into the seat cushion by the yaw. Flying is dynamic. Its monkey hear, monkey do. When we are out of trim the fabric or aluminum skin rattles and we hear air noises. We must check the nose to see whether we are skidding or slipping. We keep the nose moving smoothly in the direction of the bank, neither too slow for that amount of bank (a slip) nor too fast for that amount of bank (a skid). Flying is dynamic not static. All this information, and control input, is needed for a coordinated turn that theoretically needs no control input.

Question: shouldn't we step on the ball to trim the nose in a turn. Answer: yes if we are flying by reference to instruments; no if we are flying in contact with the earth. When flying pipeline patrol at 200' we cannot safely focus our attention inside the aircraft very much or very long. Only by getting our head up and out of the cockpit can we safely fly low level. The slip skid ball instrument is a crutch. It can help us learn to use our other senses by cross reference with it. At Contact Flying and Aircraft we move beyond the ball and reference the speed and willingness of the nose, the feel of our butt against the aircraft seat, the sound of the relative wind, and the smell of gas if we really screw it up. We use all that old pilot stuff. It's called flying.

With the exception of the gunship, helicopters do not require anti-torque pedal (rudder) coordination with the cyclic (stick). Rather the pedals are used to trim the aircraft during power changes. Some form of stability augmentation system is required on the gunship to stabilize the gun platform. These systems make them fly like fixed winged aircraft.

Dutch rolls are possible with the gunship and we may trim the nose in the turn. In normal helicopters the nose is allowed to wander in normal flight and there are no controls available to stop temporary slips and skids. Like the Ercoupe, normal helicopters have most yaw problems designed out. This makes pilot anti-torque or rudder input unnecessary in the turn.

CONTACT FLYING

While they are tremendously modern and well designed, neither the normal helicopter nor the Ercoupe provide a stable gun platform.

Because the steep but coordinated turn is so necessary in contact flying, Dutch rolls are a necessary maneuver in contact flying. Sloppy rudder usage in low level maneuvering flight is very dangerous. When doing Dutch rolls we cannot keep the distant target between our legs without proper rudder usage especially when bringing a wing back up. This funny maneuver is the basis of safe contact flying. The outcome of all contact maneuvers, now called maneuvering flight, will be in doubt until we bring rudder trim under control.

Chapter 4

THE ENERGY MANAGEMENT TURN

My first lesson at AH1-G Cobra Transition and Aerial Gunnery School at Hunter Army Airfield near Savannah Georgia was a pleasant shock. The Cobra flew like an airplane and I was required to make all turns at sixty degree or greater bank. We picked a distant target (not a heading), pitched up to lose speed, and turned aggressively allowing the nose to fall through and onto the distant target. The distant target was progressively moved closer and closer to the aircraft requiring greater pitch up and greater angles of bank to bring the sight reticule (pipper) onto the target. Flying cover for a scout helicopter a quarter mile distant from and twelve hundred feet below you required a bank of one hundred twenty degrees. We had to start slow (we circled the low bird at forty knots and twenty two pounds of torque) and get onto target fast to get the enemy's head down and to have much engagement time prior to pull up.

Pulling up from each gun run slows you down allowing a rapid return to target turn at a very great angle of bank. The slower we go the faster we turn. The pull up also gains us altitude which can be traded for airspeed in the next run. This cannot go on forever. We get lower and slower with each run eventually requiring a disengagement, power climb, and reengagement.

In the Primary Helicopter Course at Fort Walters Texas, I had not been impressed with the stability of the Hughes TH-55. Civilian models were called the Hughes 269. We called it the Mattel Messerschmitt. On my first solo flight I was to take the trainer from Dempsey Army Airfield, the world's busiest airfield at that time, to the stage field my class was using. Having many hours in airplanes I saw no problem with scratching my itching nose with my right hand. I was immediately looking straight

CONTACT FLYING

down at prickly pear cactus and rocks while the cyclic was wobbling around in counterclockwise circles. I forgot my itching nose, grabbed the cyclic, and got my ship out to the stage field.

The fat Huey (UH1-A, B or D) at the Advanced Helicopter Course at Fort Rucker Alabama was bigger and less quirky but still maintained only a general heading. So it was shocking but pleasant to get into a helicopter that flew like an airplane even if the trim was electric and artificial. It was in gun school that I first learned about energy management turns.

While it is imperative to simplify maneuvers and eliminate erratic control movement in instrument flight, at Contact Flying and Aircraft we have the flexibility to investigate wider parameters. We are also free to observe the aircraft fly hands off to better determine what its preferences are. While rigid rules save us from ourselves in instrument flight, they hurt us in contact flight. These altitude and airspeed constraints are unfortunate in the level turn, the climbing turn, and the descending turn. Eliminating this need for the level turn allows us to develop the necessary trust in our senses and the design of our aircraft to learn easily and get the greatest performance from the aircraft.

Consider the turn. Whether in a level, climbing, or descending turn the aircraft always wants to lower its nose to regain cruise as it was designed to do. By demanding the politically proper pitch attitude in these turns, we cause the student to believe this nosing down is not a desirable design trait. It is a desirable design trait and it will help us keep zoom reserve in the form of airspeed when near the ground.

At Contact Flying and Aircraft we work with the aircraft instead of against it. We demand no load turns instead of constant pitch attitude level turns. When the student at 500' begins a level turn the nose falls through to regain the same relative wind. The student naturally applies a little back pressure to keep us out of the trees and that is good for the first few turns. However, the student needs to learn that this fighting the nose up is fighting the aircraft and is wrong. He needs to learn to work with the aircraft and let the nose fall through. But how do we keep out of the trees?

Before we answer that question we need to review what Wolfgang Langewiesche teaches us about what the aircraft wants to do in <u>Stick and Rudder.</u> The aircraft likes to fly straight and level or slightly downhill. The aircraft (including its propeller) does not like to dive. The increase in relative wind is disturbing. The aircraft does not like the reduction in relative wind caused by climbing or turning and it especially does not like both at the same time. The aircraft is most comfortable with the relative wind created when airspeed remains near cruise speed and will fight the pilot who demands, for an extended period, speeds greater or lesser than cruise speed.

How then do we come to some compromise with the aircraft that will allow us to safely glean the greatest efficiency and comfort from it? How do we make no load turns and stay out of the trees? First of all we need to keep the aircraft happy. If it groans, shakes, smells of gas, feels mushy, and fights you to get its nose down; level the aircraft. Level both wings and nose. If things are not going well, let it return to its favorite flight regime.

The aircraft will climb wings level without much fuss for a short time. At Contact Flying and Aircraft we plan the turn based on what we see on the ground and where we want to go. We climb wings level prior to the turn. This is of great advantage to us because it reduces the airspeed and the centrifugal force of the turn. Another advantage is that it takes us higher above the trees (zoom reserve in the form of altitude) allowing us to use gravity to rapidly regain zoom reserve in the form of airspeed in the latter stages of the turn. Airspeed is altitude. We use cruise speed to zoom wings level from a standard five hundred feet cruise altitude to maybe six hundred feet above the trees. At this one hundred feet greater altitude and necessarily slower airspeed we can turn quickly at any needed bank to the target (the place on the ground we want to fly over). The slower we go the faster we turn. This altitude can also be traded back for airspeed as we allow the nose to fall through naturally. We do have to manage the nose tuck because the aircraft wants to regain cruise relative wind too rapidly. When we level the wings first to prevent the graveyard spiral and then the nose, the extra speed of the dive will return us back to five hundred feet while flying toward the target. Altitude is airspeed. Airspeed and altitude are two forms of the very thing that makes the airplane most happy: zoom reserve.

Therefore the no load normal turn consists of a slight climb, roll at whatever bank is necessary to acquire the target while allowing the nose to fall through, slight backpressure to manage the tuck, level the wings first then level the nose, and cruise toward the target. Climbing and descending turns are the same with slightly higher and slightly lower pitch attitudes respectively.

During no load, variable altitude (energy management) turns all the other aspects of the turn are in place. We have to initiate roll with aileron and rudder pressure then release that pressure when the desired bank is achieved. We have to manage proper nose speed of movement across the horizon with rudder. We have to manage pitch up or down as appropriate with elevator. The only difference between this turn and the level turn is the lack of load factor.

Load factor is the combined effect of centrifugal force and gravity on the aircraft when the heading and pitch attitude are simultaneously changed. Centrifugal force jumps onto gravity during curving flight (vertical and/or horizontal heading changes). While jet aircraft will power through high speed and high load factor maneuvers until the pilot passes out, small aircraft will loose zoom reserve in the form of airspeed and mush or stall.

Using the energy management turn we deal with load factor by pitching up wings level before the turn. This results in a unidirectional load and an airspeed reduction but no load factor. The airspeed reduction also reduces the centrifugal load on the aircraft in the turn that follows. By allowing the nose to find the same relative wind in the turn (fall through) we cause no load factor in the turn. We turn steeper than normal while slow to complete the turn quickly so that we don't create load factor in a turning pull up from the resulting dive. To accomplish the pull up from the dive we level the wings prior to applying back pressure to level the aircraft. This results in a unidirectional load and airspeed reduction. Thus we arrive somewhere around the original five hundred feet in cruise flight toward the new target (think target not heading). Failure to level the wings prior to pull up can result in the graveyard spiral.

Instrument flight demands mastery of the level turn (and the standard pitch attitude climbing and descending turns). Yet when we demand the level turn we severely limit the safe maneuverability of the aircraft. Now load factors must be stressed and angles of bank must be limited. Personal minimums become important. The objective becomes to fly rigidly with wings and nose near level at all times.

While this rigidity helps our student transition to instrument flight, it limits his comfort and confidence in contact flight. And that lack of confidence is well founded, for when he must turn sharply and let the nose fall through to maintain necessary airspeed for flight he will be unable to accomplish this. Instead he will maintain wings near level and plow straight into the obstacle or stall the aircraft attempting to climb over it. Given normal Private and Commercial flight training today, our students rightly fear what might happen if they someday really need to move quickly and outside the box they are trained in.

Contact Flying and Aircraft emphasizes the energy management turn. Because no load factor is created, this is the only really safe turn in low level contact flight. Takeoff and landing are low level ground reference maneuvers, even if we are lulled into feeling secure with the high altitude pattern. Pattern work can be safer if we allow the nose to fall through naturally in all turns. I honestly cannot teach normal patterns. After my student fights that little training aircraft up to within four hundred feet of pattern altitude, I cannot condone the PTS required climbing turn on crosswind. I simply must have the nose fall through at least enough the get the stall warning devise to cease warning. I cannot understand the PTS required standard base to final high load turn (not allowing the nose to fall through) with a tailwind. At Contact Flying and Aircraft we teach our students to take advantage of natural forces and turn downwind on crosswind so that base to final can be upwind at the slowest possible groundspeed. But if we have to make arbitrary left or right turns, we should at least instruct them to let the nose fall through if they overshoot final. Contact flying allows us to better mesh with our environment and take advantage of energy savings. It is safe, fun, and economical.

In helicopters the energy management turn involves a cyclic climb, a turn to target, and allowing the nose to fall through naturally. Power changes in helicopters are problematic and require aggressive anti-torque pedal management. Therefore we avoid using collective pitch (power) to climb unless we need to climb significantly as in climb out from a gun run. Here we use max power (40 pounds of torque in a Cobra) but this in not necessary in a normal or steep turn where we simply make a cyclic climb to slow for the energy management turn.

The energy management gun run, like the crop duster turn, requires maximum use of all controls. In the killer (Cobra) we orbit the hunter (scout) at 1200' above ground level (AGL) nose well up at forty knots and twenty two pounds of torque (the best power setting to achieve a sixty degree pitch down dive). When the scout calls "taking fire," we immediately roll into the degree of bank (usually 90 to 120 degrees) that will most rapidly align our nose with the target. We level the tip path plane (wings) check trim (rockets will immediately weathervane into the relative wind upon release from the wing stores) and fire as many pairs of 2.75 inch folding fin aerial rockets as possible before pull up is necessary. We want to pull up prior to overflying the target and we must pull up prior to reaching the VNE (Velocity Never Exceed) of 190 knots. At this point we pull full power with the collective control and pitch up radically with the cyclic control (stick) keeping the tip path plane (wings) level until the airspeed dissipates to near forty knots. Now we roll into a 120 degree bank and start the cycle all over again. We continue making runs until we get too low and too slow. At this point we must disengage and climb back to 1200' AGL to reengage.

Chapter 5

THE APPARENT RATE OF CLOSURE APPROACH

At Ag Flight, a crop dusting school in Bainbridge Georgia, I was given charge of the zero time students who came by twos on Monday every other week from October to March. We had an old Army Air Force hanger on a nearly abandoned airport and about twenty tail wheel aircraft including four Pawnee agricultural airplanes and an Air Tractor agricultural airplane. We averaged about twenty students. Half of the students were zero timers while the other half held Private Pilot or Commercial Pilot Certificates. We preferred students without any formal experience because less deprogramming was necessary.

There are advantages and disadvantages to a contact only (no integration of instruments) primary training program. At Bainbridge we were the only operation on the field other than one charter business. We had multiple runways over a 640 acre area. Three runways were paved and one was grass. With a four hundred foot pattern altitude and the apparent rate of closure approach we could practice ten takeoffs and landing each hour. Using ground reference only (the airspeed indicator was covered) students could learn quickly. This was because at low altitude using ground reference they could receive immediate knowledge of the results of their control inputs. Needle, ball, airspeed, artificial horizon, directional gyro, altimeter, and vertical speed indications inside the cockpit at altitude pale in comparison with the immediate and accurate information garnered from ground reference combined with relative wind and engine sounds, gas smells, feel of the relation of the stick to ones stomach, feel of pitch, roll, and yaw forces through ones behind, and yes even the taste of acid reflex or absence thereof depending on the quality of the flying. The only disadvantage of contact flying was that our students required separate and conflicting training in order to pass the Private Pilot and Commercial Pilot flight tests.

At Contact Flying and Aircraft we get our young crop dusting students up at "zero dark thirty" (before sunup) because they will have to get up before the sun to be successful crop dusters. We teach them to cook and clean because crop dusters need a second career to keep from starving in the off season (October to March). We require every plane to be pulled out of the big hanger and propped or cranked each weekday morning because young people tend to stay up too late and sleep too late to be successful crop dusters. If they have to pull all the airplanes out and crank them they will usually fly them and complete the course on time.

All of our students, using the apparent rate of closure technique, solo by Wednesday of their first week there. All three of their first solo landings are either on the numbers or short of the numbers toward the approach end of the runway. This is possible because they are flying solely by contact with the earth. Chasing airspeed is neither taught nor allowed. This is neither strange nor new. In my youth I remember that flight training through solo was similarly rapid and precise.

If we land by reference to airspeed we are flying by reference to instruments. It is difficult, using instruments as our main reference, to land on airports or landing zones with less than three thousand feet of landing area available. With contact students, especially those who desire to stay with bush, crop dusting, or light sport flying, I teach the brisk walk apparent rate of closure approach. This Army helicopter training technique stabilizes an approach based upon the relative appearance of things on the ground rather than upon numbers on gauges. It utilizes Mr. Langewiesche's "stall down" procedure. By keeping the runway number or any object in the desired touchdown zone closing with us or coming toward us at a steady apparent brisk walk rate of closure, we will arrive at the touchdown point with no zoom reserve either in airspeed or altitude.

The beauty of this approach is that we have plenty of airspeed out over the final approach obstructions, yet we lose this extra airspeed by the time we need to touch down on the numbers. The optical use of the apparent fast walk closure rate causes a continuous gradual decent and deceleration to touchdown. *See Figure 10.* This results in a spot landing every time.

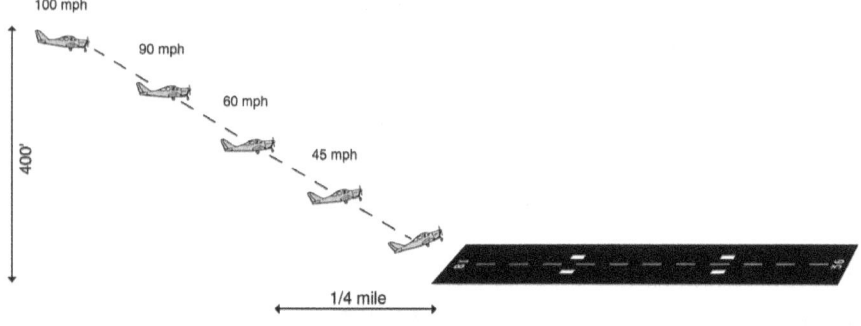

Figure 10

The apparent brisk walk rate of closure approach will work with any approach speed until about one quarter mile out where things begin to appear to speed up. In the figure, estimated airspeeds are indicated only as a contrast to normal stabilized airspeed approaches. The airspeed indicator should not be used on the apparent brisk walk rate of closure approach because it is not relevant. At altitudes above five hundred feet everything appears to move no faster than a brisk walk. We must be low enough on final to see faster apparent motion. Ground reference maneuvers, including landing, become increasingly difficult to learn as altitude is gained. This is because a successful outcome to ground reference maneuvers depends upon observation of ones groundspeed not ones airspeed. Above five hundred feet, ground reference maneuvers become rote and we find the student referring to gauges rather than to the ground. Reference to gauges when near the ground is unnecessary and dangerous.

While the brisk walk apparent rate of closure approach was designed for helicopters it works quite well for airplanes as well. With so many different designs coming out in light sport aviation, there might be an airplane with a chin bubble. In airplanes we seldom have a chin bubble to look through. If a chin bubble becomes available in light sport aircraft we must not use it during the apparent rate of closure approach or we will get too slow on very short final. With most airplanes this is not a problem because as you get very close to touchdown you have to pick a new aiming point further down the runway. The old aiming point is blocked by the nose. This prevents our actually coming to a stop like the helicopter. In a fixed wing aircraft this would result in a stall on short final.

We will be behind the power curve on very short final. This causes an inverse relationship of controls. At these very slow speeds with full flaps and quite a bit of power, pushing the nose down will cause assent and pulling the nose up will cause descent. The throttle must be increased as pitch attitude is increased to slow down and maintain the brisk walk. If a downdraft or downwind shear is encountered full throttle must be applied to arrest the sink then adjusted. If an updraft is encountered the throttle must be closed until a change is noted then adjusted. While easy enough to grasp as a first learned technique, the apparent rate of closure approach may be too difficult a transition for those who have landed "normally" for some time.

Landing from low altitude is of great advantage to the new learner. It is extremely difficult to get a light aircraft down accurately from one thousand feet above the ground. If altitude is airspeed and we demand one thousand feet of it plus a fence speed of 1.3 times Vso (Velocity of Stall with everything Out), how do we expect anyone to get down on the numbers? This problem has evolved over time but stems from the objective of today's landing theory which is to simulate the instrument approach on all contact approaches.

As corporate and airline aircraft grew in size and complexity, the traffic pattern for all aircraft moved steadily upward from six hundred feet to one thousand feet. This mandated 1.3 Vso and 1000' starting altitude simulates the ILS approach. Today's student flying the same aircraft I flew

CONTACT FLYING

fifty years ago is expected to shoot a simulated ILS approach when he should be learning contact flying. On page 148 of <u>Flight Instructor's Pocket Companion</u>, we are instructed to "Pick your landing spot about a thousand feet down the runway from the end. Do not try to land right on the end except when the runway is very short." And on page 153 we are told, "The best approach speed is 1.3 times the stall speed. For a short field landing, it can be reduced to 1.2 times stall speed in the final stages of the approach."

While justified for students going on to corporate or airline work, these mandates work against the proper training of the other 95 percent of the student population who will continue to fly small aircraft. And if we end up spraying out of a 1500' strip or fly hunters into 800' unimproved landing sites, 1.2 times stall speed on short final will give us more trouble than a go around can handle. The use of an approach that is meant to simulate going down the glide slope on the ILS is just not appropriate to most small aircraft contact approach situations. On the common 3,000' Midwest airport, brakes will be required to finish any approach using 1.3 Vso over the fence.

The brisk walk apparent rate of closure is dynamic. On short final airspeed and pitch attitude are constantly changing. We cannot fly this approach with set or stabilized power. We must dynamically apply more or less elevator to maintain a brisk walk closure rate and more or less throttle to maintain either glide slope or a 5' constant altitude. We must use the glide slope angle that will both clear obstructions and allow touchdown on the numbers. To accomplish this, we descend until we can see just over the obstruction to the numbers. If there are no obstructions, descent to 5' is appropriate. Once this minimum glide slope angle is reached, we put sufficient power back in to maintain that angle. I don't tell students to adjust the throttle here. Throttle is our least responsive control. In low powered aircraft it is necessary to make dynamic and radical power changes to have sufficient control of altitude at low speed. When too high on glide slope, we need to get the throttle completely closed and then adjust as necessary. When we hit a downdraft we need to push the throttle in all the way then adjust as necessary. If we climb too much we can reduce it. We're better when we move.

This brings us to another training dogma problem similar to the erroneous idea that a downwind turn should be no problem to a crop duster. Many training manuals and articles say we have to control glide slope with elevator and speed with power (they said just the opposite fifty years ago). The elevator will elevate and extra power will bring extra speed, if we come all the way down the glide slope at 1.3 times stall speed. The problem is that light airplanes don't land well at 1.3 times stall speed. And when we finally slow down enough to hold off, the elevator no longer elevates but just the opposite. This is confusing to the student who is taught little about behind the power curve operations.

Contact Flying and Aircraft students understand that when they have used up their elevator to the point that further back pressure will result in mushing, they will have to use throttle to arrest the sink. Contact students need to understand that slowing and mushing are desirable short final landing traits. Slowing allows us to land on the surface available with least harm to gear, tires, and brakes. Slowing prevents destruction of tail wheel aircraft if a ground loop occurs. Mushing allows us to add power and force extra air over the wing with the prop. This extra air will allow the wing to fly even slower. Flaps allow us to go even slower. Slats allow us to go slower. Swinging the wing (helicopter) allows us to go much slower. The slower we go the better we land. The beauty of the helicopter is that we can put it down in a very tight area (hover hole) because we can zero out the airspeed. This is possible in small fixed wing aircraft only if the headwind equals or exceeds the stall speed.

We don't want to stall at one hundred feet above the ground but we want to stall at just an inch or so above the ground. The apparent rate of closure approach will cause us to reduce airspeed and altitude steadily until touchdown. As we begin to significantly lose buoyancy on short

CONTACT FLYING

final we should start to fall through requiring throttle, not elevator, to arrest the rapid sink rate. Power is also required if, while at low zoom reserve in the form of airspeed, we experience a sink due to a downwind shear or a downdraft. In either case application of up elevator will always cause a more rapid sink rate (exactly as it did in the Vx takeoff scenario.) I say again, application of up elevator will cause a more rapid sink rate. We arrest the sink with power. If the ship sinks below glide slope, contact pilots add power to stay on glide slope.

I am not trying to confuse or mislead anyone with this radical but effective approach procedure. It should not be attempted without the help of someone familiar with the apparent rate of closure approach. We do not drag it in at one slow speed because that procedure leaves you hanging on the prop over the obstructions and does not result in a gradual deceleration to touchdown. We do go behind the power curve just prier to touchdown (at or even below stall speed.) We are not in position, on those approaches without obstructions, to make the field if the engine fails. Small Continental engines have a tendency to hesitate when the throttle is handled aggressively. This apparent rate of closure approach is not for everyone but will certainly help those needing to land on short and or unimproved strips. The knowledge of this technique, however, helps any of us understand light aircraft better and affords tremendous speed control when needed. Only the use of this technique will bring tail wheel insurance prices down because the 1.3 Vso approach will destroy them every time in a fast ground loop.

If landing is a ground reference maneuver and if groundspeed is important in ground reference maneuvers, then it follows that wind direction must be considered in the execution of good landings. If turning to miss objects and align with objects on the ground is truly easier at lower verses higher groundspeeds, then it follows that arbitrary directions of turn in the pattern with no reference to wind direction are neither easier nor safer turns. When landing with a strong crosswind, we should make base to final turns on an upwind base. If doing practice takeoff and landings at the same airport, it would make good wind energy management sense to make our crosswind leg in a downwind direction when winds cut across the runway we are using. This downwind crosswind leg would cause us to make our base to final leg into the wind at our slowest groundspeed. *See Figure 11.*

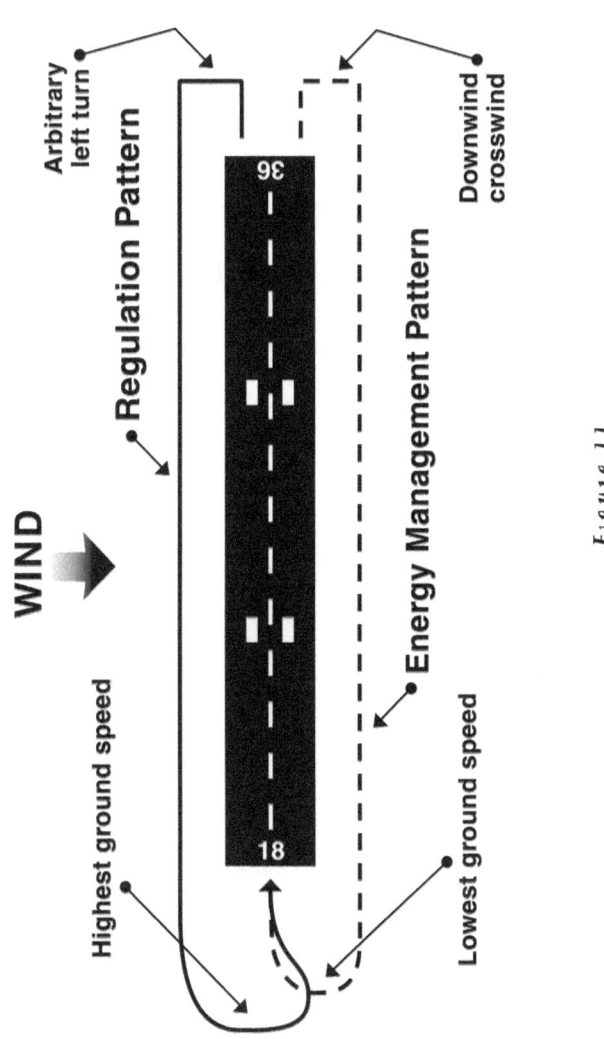

Figure 11

Our need for standardization and regulation causes many less safe downwind base legs at our fastest groundspeed. See illustration. It also inadvertently precludes effective crosswind training in moderate crosswind conditions. Large and obvious conditions like a moderate to strong crosswind are more useful in early training than are subtle things like a light crosswind. It is much easier to teach, and more self confidence building to learn, crosswind takeoff and landing operations in moderate to strong crosswinds. If we are so constrained by official suggestion that we cannot allow a student to make right traffic at a left traffic non-controlled airport, we are unable to give him the confidence building satisfaction of handling tough crosswinds during pre-solo. Now he has a confidence problem: what if the wind comes up on his first solo?

Making left base to runway eighteen in *Figure 11* with strong easterly crosswinds goes against good ground reference wind intelligence. We don't punish ourselves that way when spraying even in light winds. When we spray crops we always try to work crosswind and move one swath width upwind after each turnaround. *See Figure 12.*

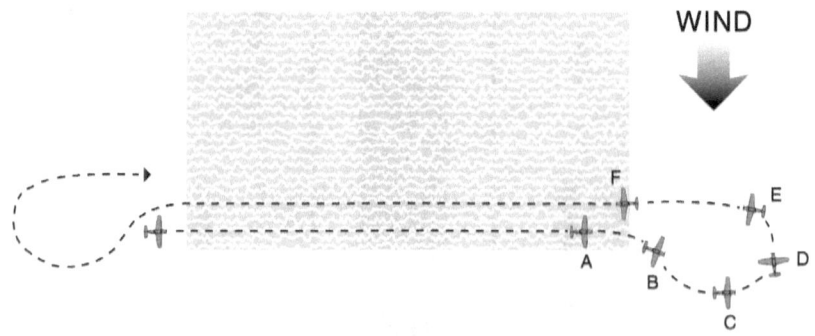

Figure 12

Using ground effect in the field (A) we have energy to zoom over the obstructions at the end of the field (A). As soon as we have the obstruction cleared (we teach "clear the wing") we turn downwind to an appropriate angle (based on wind direction and speed which is determined by comparing how ones behind crosses the earth with where the nose is pointed) (B) away from the row we wish to return to. This downwind first or "fall off the target downwind" is critical to any ground reference maneuver: crop duster turn, gunship return to target turn, takeoff and landing pattern work crosswind turn, lazy eight the way we used to teach them, weather check return to runway you just left without doing a pattern, and what now is now called maneuvering flight. With wings level (C) we climb as much as load and conditions will allow slowing the aircraft (the slower we go the faster we turn). Now (D) we must bank steeply (sixty to ninety degrees) to complete the 180 degree turn and return to the next crop row forty feet upwind from the one we left (F). It is critical to bank steeply enough (allowing the nose to fall through) so that we can level the wings (E) before pulling the nose up as we enter the field (F).

It is absolutely crucial in this crop duster turn or the gunship return to target to climb wings level (if we climb and turn we create load factor), to turn aggressively allowing the nose to fall through naturally to prevent load factors at such a steep bank, and to level the wings before pulling up to prevent the graveyard spiral. If we attempt a medium or shallow bank we will have to rack it around and pull up at the same time, creating load factors and the graveyard spiral, to get back into the row. The steep bank is absolutely necessary to return to the target within sight of the target. Allowing the nose to fall through is absolutely necessary to prevent the high load factor mush or stall. We hold a little back pressure as we allow the nose to fall through naturally because the aircraft will try to return to cruise relative wind too quickly and thereby loose excessive altitude. Once headed toward the field (E to F) nose pitch adjustment (with the wings now level) may be made to clear obstructions. This pitch adjustment and the pull up in the field will load the aircraft. With wings level, however, no load factors are involved.

The crop duster turns downwind after pull up so that when on base to the next swath run, he now has a headwind, which will slow his groundspeed. Now he has time to line up on the correct upwind row (F) at his slowest groundspeed (E). When we consider our ground track, nature has increased our turning rate free of any extra banking on our part. With slower airspeed our rate of turn increases but the extra bonus is that with slower groundspeed our path across the earth is deflected toward the target weather it be crop row, enemy, or runway.

Why can't this same technique be modified to allow good energy and groundspeed management in takeoff and landing operations? At uncontrolled fields, could not the regulation state, "If winds are observed to differ more than ten degrees from runway heading, the crosswind leg will be flown downwind and the base leg will be flown upwind?" *See Figure 13.*

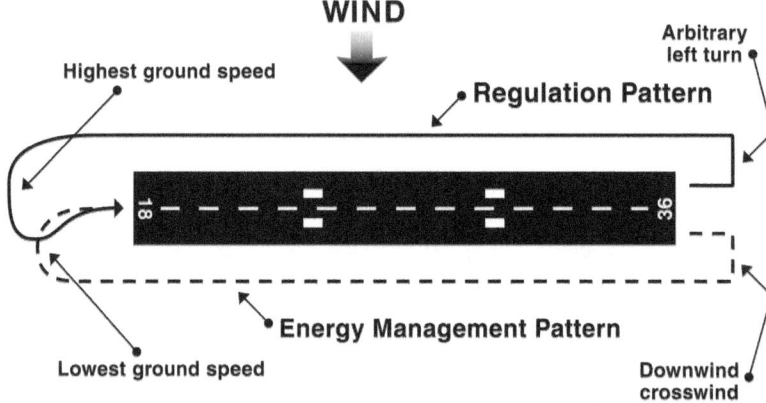

Figure 13

This would cause right base to final on runway 18 to be upwind at the slowest groundspeed and allow our student time to make the turn to final, get lined up with the centerline, and assess the amount of rudder needed to hold the centerline or modified centerline and assess the amount of bank necessary to hold the aircraft in a lateral position from which he could make the runway. In contact work we endeavor to always fall off the target downwind allowing an upwind return to target.

Using the brisk walk apparent rate of closure approach, there is no need to memorize different landing speeds for various headwinds and gust spreads. Our airspeed will increase exactly the amount of the headwind while our groundspeed will be and appear to be the same brisk walk as in the no wind approach. Here is the true stabilized approach for light aircraft. The brisk walk rate, at which our aircraft appears to approach the runway, is the same regardless of conditions.

One cautionary note; in a strong headwind the aircraft may actually stop (groundspeed zero) and hover above the runway. A very low pitch attitude might be necessary to maintain the brisk walk, just prior to touchdown, and we don't want to touchdown nose gear first. While this may be disconcerting at first, exposure will lessen the discomfort. We can reduce power to touchdown or simply hover-taxi to just short of the desired runway intersection and then reduce power to touchdown. We can raise the nose slightly just at touchdown. If we raise the nose too soon we may drift backwards slightly.

Flaps are wonderful inventions that really help the airplane land slower, which is the whole point of the landing. Flaps give the aircraft a lower pitch attitude without increasing airspeed. We can see over the nose to better control the aircraft. Flaps allow slower flight with more stability than flight at the same airspeed without them. At Contact Flying and Aircraft we teach contact students to use full flaps on all landings.

In strong crosswinds the extra control and speed reduction provided by full flaps allow us to land the ship at an angle to the runway when we have run out of rudder. Mountain and desert airports, where afternoon winds are strongest, tend to have but one runway. This single runway, however, is often long and wide. In this situation (*See Figure 14.*) we can line up on final to determine the amount of left bank and right rudder required to hold the nose straight down the centerline (with experience we will know we have insufficient rudder to use the painted centerline).

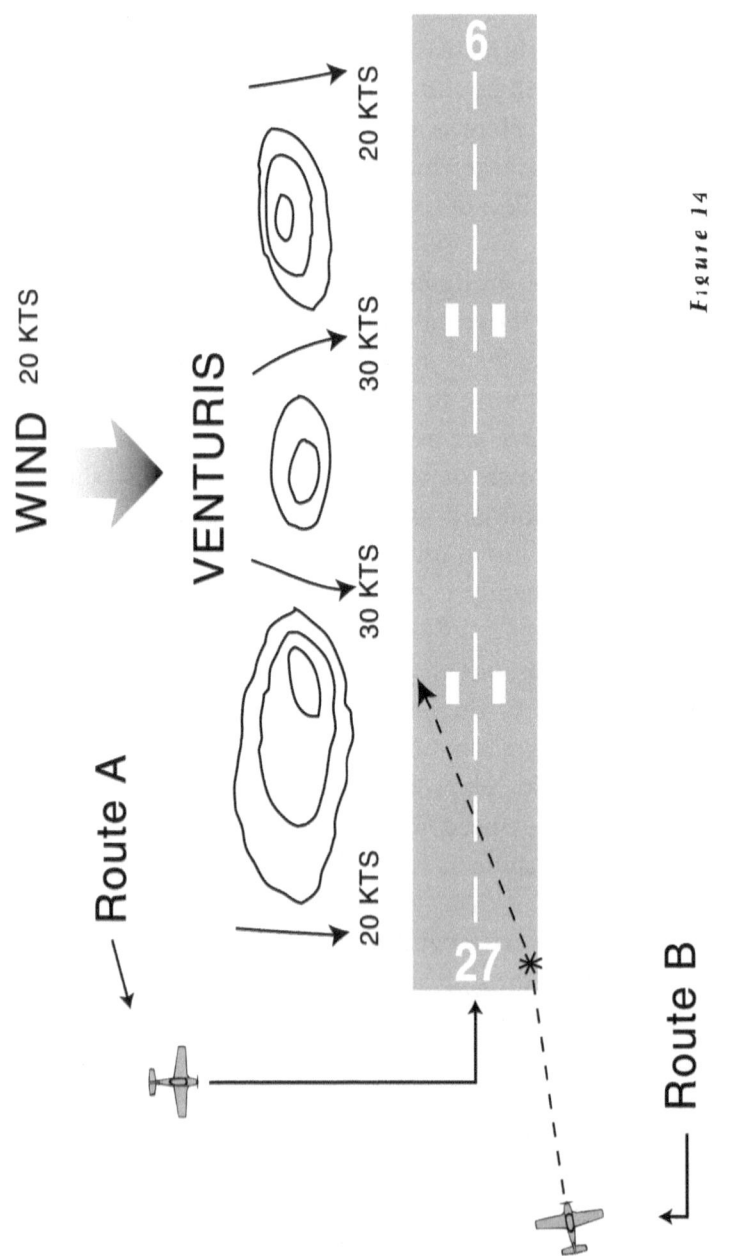
Figure 14

Given these conditions and present training technique limitations, we would either not be able to hold the centerline (we would run out of rudder) or not be able to remain over the touchdown zone. Yet we have been flying and even training in these conditions in the mountains since aviation began. The politically correct school solution would be to not have gone. That attitude is not helpful when one is there and it is many miles across the desert to another airport with fuel.

Let's look not at what ought to be but at what is out there in aviation every day. This mountain and desert situation is not unusual in the West. Any of us may someday have to deal with it. While maintaining a brisk walk apparent rate of closure and full flaps, we make right base to a modified centerline optically created from the right fore corner of runway 27 and the left big airplane aiming mark. *See Figure 14.* Using full flaps and the brisk walk apparent rate of closure we can safely land on the one thousand feet of runway available. Increased crosswind and gust spreads may cause us to increase our airspeed (this is automatic if we are approaching at an apparent brisk walk) but our groundspeed will stay the same. We must be ready to use the throttle aggressively in unstable air. We must hold the left wing down through touchdown and rollout and taxi. We must hold the right rudder to the stop. If the nose comes left off the modified centerline we must be prepared to move the far end of the modified centerline back toward the number 27 a bit. Landing liaison Cub, Aeronca, Cessna, and Taylorcraft airplanes perpendicular across Army Air Force runways was not uncommon in an age of artful flying.

If we are used to landing short and hover-taxiing to the desired intersection, we will have no problem with this strong crosswind landing. If we have made thousands of long landings using half flaps, we will have a confidence problem here. After touching down on the NE corner of Runway 27 and rolling out a bit we will slow sufficiently to turn down the runway at taxi speed at the big airplane aiming mark. If we have approached to the right fore corner of Runway 27 at an apparent brisk walk, there will be no need to apply brakes.

CONTACT FLYING

The landing accounts for the greatest number of training accidents. The AOPA Air Safety Foundation study of instructional accidents from 1992 to 2001 found that 33.1% of all dual accidents and 50.6% of all solo accidents were landing accidents. While takeoff and go around account for more than 20% of the instructional fatalities, landing only accounts for 12.1%. Taking off too slow without first establishing zoom reserve in the form of airspeed is the big killer in training. Landing too fast, however, eats the most tin and skin.

But the landing is the greatest confidence builder in the flying game. I cannot over stress its importance to student confidence. Our students must believe that they can land by themselves before they solo. Delay of solo to get extra practice will not inspire this confidence. Only early success resulting from rapid transition from limited stick time to total stick time will inspire them. If my student has confidence in my flying skills, I have failed him. He must have insufficient data upon which to judge my flying abilities or I have stolen enough of his stick time to damage his confidence. For him to be comfortable with my teaching abilities is sufficient. How long could a golf pro survive if he forced his students to watch him hit balls?

With my contact students I talk constantly but seldom touch the controls. Using energy management, a four hundred feet pattern, and the apparent rate of closure approach we can get ten takeoffs and landings each hour and a solo in six to ten hours. Skip this paragraph if you get bored but a typical circuit in a tail wheel airplane goes something like this: *"Give it some power. Keep the yellow taxi line between your legs. Reduce power. Keep your left foot on the left side of the line and your right foot on the right side of the line. Push left then right rudder to the floor. Push them to the stop like this. Ok! Good! Left, right, left, right. Walk the rudders. Pull the throttle all the way back. Walk the rudders. Use full left rudder and power to bring the tail around. Pull the throttle back to the stop. Walk the rudders.*

Line the centerline up between your legs. Walk the rudders to the stops. Smoothly add full power. Anticipate left turn. Walk the rudders but favor right a little. As we gain speed we will need less rudder movement and more rudder pressure. Walk the rudders a lot. Keep the line between your legs. When the tail feels light push it up smartly to level the aircraft. Anticipate precession and favor the right rudder. Walk the rudders. Wiggle the stick fore and aft to find the nose level best acceleration attitude. See how she accelerates with the tail up. Listen to and feel the aircraft. We're light; pull her off. Anticipate P-factor by walking the rudder but favoring right rudder. Push the nose over to stay in ground effect. Push the nose down. Walk the rudders. Leave the aileron alone. Wiggle the stick fore and aft. Look at the bottom of the tree. Walk the rudders. Wiggle the stick. Stay at five feet. We feel zoomy. Look at the top of the tree. Fly over the tree. Right rudder. Wings level. Nose level. Clear the wing and pick a point under the wing to fly to. Turn to that point. Let the nose fall through. Level the wing. Climb a bit. Clear downwind leg wing and pick a point to turn to. Turn to the point. Let the nose fall through. Cruise power. Check the trim control and set for cruise. Keep the point between your legs. Abeam the numbers reduce power somewhat and lower the flaps some. Clear and pick target under base leg wing. Make a descending turn to that target. Let the nose fall through. Get ready for the turn to final. Get us over the centerline extended. Turn final. Let the nose fall onto the centerline. Put the centerline between your legs and keep it there with rudder not aileron. Full flaps. Keep us lined up with rudder not aileron. Walk the rudders a little. Lock the aileron. Look over the tree to the numbers. We see too much grass between the tree and the runway; pull the throttle all the way back and adjust when we settle enough that just a bit of grass is visible. Bring the nose up a bit to test the amount of zoom reserve in airspeed. We're climbing, reduce power. Maintain the glide path to the numbers with power. Maintain an apparent brisk walk toward the numbers with elevator. We're sinking and losing grass before the runway; add power. Now reduce power we see enough grass. Leave the nose down. We correct the sink with power. Lock the aileron. We make all lateral adjustments with rudder only. Too fast an apparent rate of closure; bring the nose up a bit with elevator. Too much grass now; power back a little. Check height out side window. Walk the rudders. Wiggle the stick. We're falling through; add a little power. Leave nose down. Walk rudders. Level the wings with rudder. No aileron. Power off. Walk the rudders. Look on down the runway. Hold it off. As main wheels touch, push the nose over to level the aircraft. Walk the rudders a little then a lot as she slows down. Wiggle stick fore and aft to keep her level. When the tail wants to sag, lock it down with full aft stick for tail wheel control. No brakes. Walk the rudders a lot now and notice that the pressure is little. Walk the rudders to the stop and taxi back."

I broke my own no brake rule with a student in a Luscombe 8A. Intelligent instructors put the student in the right seat (the 8A has no brakes on the right side). I put most students in the right seat as well. Most tail wheel transition students will go on to fly other tail wheel aircraft and most tail wheel aircraft are tandem and thus have the throttle on the pilots left anyway. This student wanted to stay with the Luscombe so I started him in the left seat and told him not to use the brakes but to turn the engine off with the mag switch if worried about hitting something.

We were training out of a 3/8 mile grass crop duster strip in the middle of a soybean field. The student was not slowing sufficiently on the approach and by the time I took control we were on the grass with the tail up and excessive speed. I was not worried about an overrun at low speed into the tall grass of the next field, but I was unsure about clearance on the pickup parked down there. I cheated as far right as the narrow strip would allow and finally got scared and called for brakes. The student had never used them because I had not allowed it. Somewhere in the braking or in my cheating right we got the right main into the soybeans. I had enough rudder control to hold her straight but could not get back to the grass. We entered the soybeans at a very slight angle, slowed rapidly, and went over on our back. Bent airplane (the student did switch the mags off saving the engine); bent pride!

The trike solved a lot of our landing problems but even the tricycle gear cannot eliminate problems with extremely high touchdown speeds. Current landing theory uses Mr. Langwieche's "hold off" technique. To simulate the "stabilized approach" (contact ILS approach) the PTS demands a stabilized airspeed of 1.3 Vso, which makes the stall down by definition impossible. Further, the PTS demands a 1000' pattern that contributes to very long landings using the hold off technique. By demanding too much speed to stall down (mush down) and too much altitude to run the ship in very low on short final using the hold off, the PTS has set the small aircraft pilot up for a go around.

Chapter 6

THE GO AROUND

I was the flight instructor for the Civil Air Patrol during my family's fourteen years on the Navajo Reservation in the 70s and 80s. Most members of this Air Force Auxiliary flying club were middle aged and worked professional jobs. Not only did they strictly adhere to FAA and CAP (Civil Air Patrol) rules and regulations but they always interpreted those rules and regulations in ways that were least favorable to the ability of the small club to fly the one Cessna 172 given us. No matter how much I emphasized the greater power of natural forces in the mountains around Tohatchi, I knew in my heart that most of my students would probably stall the aircraft rather than go below the 500' minimum altitude above manmade objects in what was then called uncontrolled airspace. Looking back I am grateful that the go around was not so highly encouraged in those days. My CAP students might have considered that a mandate and might never have landed. Sometimes I exaggerate too much.

According to AOPA's study of instruction accidents from 1992 to 2001, 5.7% of dual fatalities (# of accidents) and 6.7% of solo fatalities were on go around. These numbers may seem small and acceptable unless we look at their causes and see how we might reduce the need to go around. I believe the Practical Test Standard 1.3 Vso on landing and Vx or Vy on takeoff cause our students to believe that fast speeds are necessary on landing and slow speeds are necessary on takeoff when just the opposite makes more common sense. I believe approaching slower, whether or not we use the brisk walk apparent rate of closure, reduces the need to go around.

I do not emphasize the go around in training students because I teach the brisk walk apparent rate of closure approach. This approach puts all landings on the numbers which is a tremendous safety advantage over the "stabilized approach" technique. A brisk walk apparent rate of closure will dissipate zoom reserve (both in altitude and airspeed), so that the go around can safely be initiated at any point during the approach. However, if we initiate the approach from 1,000' AGL, plan to arrive at the large aircraft touchdown zone at 1.3 Vso, rotate, and hold the ship off as long as possible, we will be touching down some 2,000' down the runway. At this point (2,000' down a 3,000' runway) a go around is no longer a safe option. To be safe the go around would have been initiated before passing the large aircraft touchdown zone. This means that a safe go around is just not likely at a majority of airports in this country when using a stabilized approach to the large airplane touchdown zone.

A token 1.2 Vso is not the solution to this problem. We cannot land either on the numbers or the large aircraft touchdown zone at any speed significantly above stall speed. We cannot land in a safe distance (1,000 feet or less) down a normal 3,000' runway after going down the glide slope so fast as to be able to control glide slope with elevator. On the NDB, VOR, or GPS approach to small airports, upon making visual contact we have to rapidly decelerate below 1.3 Vso and rapidly descend below glide slope to get stopped on the runway available. Unless we have dissipated all extra speed and altitude by the time we cross the fence, we will go long enough to require maximum braking.

I don't dislike the go around as much as the mindset that goes with it. I feel that going around is over emphasized as a solution to technique problems. Students are encouraged, on page 157 of <u>The Flight Instructor's Pocket Companion</u>, to go around. They are there admonished that "the pilot who goes around will not be belittled or criticized by the real professionals, but will be praised instead." This leads the student to believe that landing long is quite normal. Further it implies that using the contact ILS technique rather than learning to get it down on the runway, wherever you are, is desirable. The contact ILS technique assumes the availability of long runways. For the student who trains only at large airports and goes on to corporate or airline work, short runways are not a problem.

The problem arises when the student or even the ATP without recent short field experience arrives at Smallville in a small airplane to visit family. At 1,000' up and one half mile out to a 2,500' runway things already are not right. To be safe a go around should be initiated at this point. The attempt to lose so much altitude while holding a 3% glide slope and 1.2 Vso simply will not work. The extra 400' above what patterns used to be will translate into excessive speed over the fence. The 1.2 required minimum speed with extra speed added for gust spread will translate into excessive speed over the fence. These excessive speeds from two arbitrary sources will result in the aircraft going excessively far down the runway before touchdown. What will happen at Smallville is a very dangerous go around initiated from 2,000' down a 2,500' runway.

If he survives the go around he will drag it in at less than 1.2 Vso from one half mile out. This dragging it in or using a very slow approach speed from very far out is dangerous but a pilot trained only to land at large airports using large airplane approach techniques has no other choice.

I have witnessed many times a similar problem resulting from the lack of contact flying skills. In the western states small aircraft often arrived at a long single runway with insufficient fuel to go another 150 miles to the next airfield with services. In the afternoon the winds were almost always strong and often were directly crosswind. The pilot attempted several crosswind landings but ran out of rudder each time and made a go around. On occasion this resulted in bent metal and hurt pride. This was followed up with cool reactions from other pilots and letters of proposed civil sanctions or penalties from the FAA. When something gets bent and many at the airport are talking about what piece of data the pilot didn't know or didn't check we instructors need to lessen the tension not heighten it. Sometimes, fair knight, the dragon wins.

I certainly do not fault the pilot. He was led to believe that proper scrutiny of all available weather and airport data, the POH, etc. would have prevented this incident. But he had done all that. Shockingly, in the real world, preflight planning and the go around will not fix everything. This shock is not the result of poor judgment or lack of ability.

Nor do I fault the instructor. He just taught what was in the book like he had been taught. The fault lies with a training system so regimented and regulated it cannot conceive of reality. What is actually happening out there is dealt with from one point of view only. Who violated regulations and how are we going to get them? The only bureaucracy that looks inward for the source of a problem is the Army, and that is in time of war. Flying contact approaches as if they were instrument approaches causes reliance on the go around. Statistically the go around is dangerous.

I understand that there will be times when we must go around. Someone may taxi onto the runway or the tower may request a go around, etc. I believe that the go around can be flown more safely if we emphasize speed rather than altitude. If we achieve zoom reserve in the form of airspeed prior to attempting a climb, we will be able to maneuver around the problem without mushing or stalling. Climbing is not always the safest way to avoid problems.

Chapter 7

THE DOWNWIND TURN

In 1995 I sprayed most of the irrigated corn in Furnas County Nebraska four times (root worm, two broods of corn borer, and spider mites) for a total of twenty three thousand acres in three months. This was pretty good for one man and a Pawnee. Actually it was one man and three of the best Pawnees I have ever flown. Mr. Jerry Benhke, my boss, had the best run Pawnee operation in the country. He had a spreader plane, a poison plane, and a herbicide plane. I owned a pretty good Pawnee in the early 80s but any one of these planes was the best spray plane I had ever been in. Mr. Benhke was the FAA's mechanic of the year in 1957 and he owned the cleanest planes and the cleanest crop dusting operation in the country.

One of the fields I sprayed and treated (spread dry material with the spreader) illustrates the downwind turn problem that is very much a problem for low level contact flying. This eighty acre rectangular field was 1/2 mile long north to south by 1/4 mile wide east to west. *See Figure 15.*

The two possible ways to treat this field was crosswind east to west or upwind and downwind north to south. The wind was fifteen knots from the south. We generally spray or spread crosswind starting on the downwind border of a field, but I wasn't about to make twice as many 1/4 mile passes with Pounce granulated insecticide and waste ten percent of this expensive material.

I treated this field north to south so that there would be 37 swaths. 1/4 mile is 1320 feet. The swath width for a Pawnee is 36 feet. 1320 feet divided by 36 feet is 37 spreader swaths. If we made our swath runs east to west, this would require twice as many swath runs.

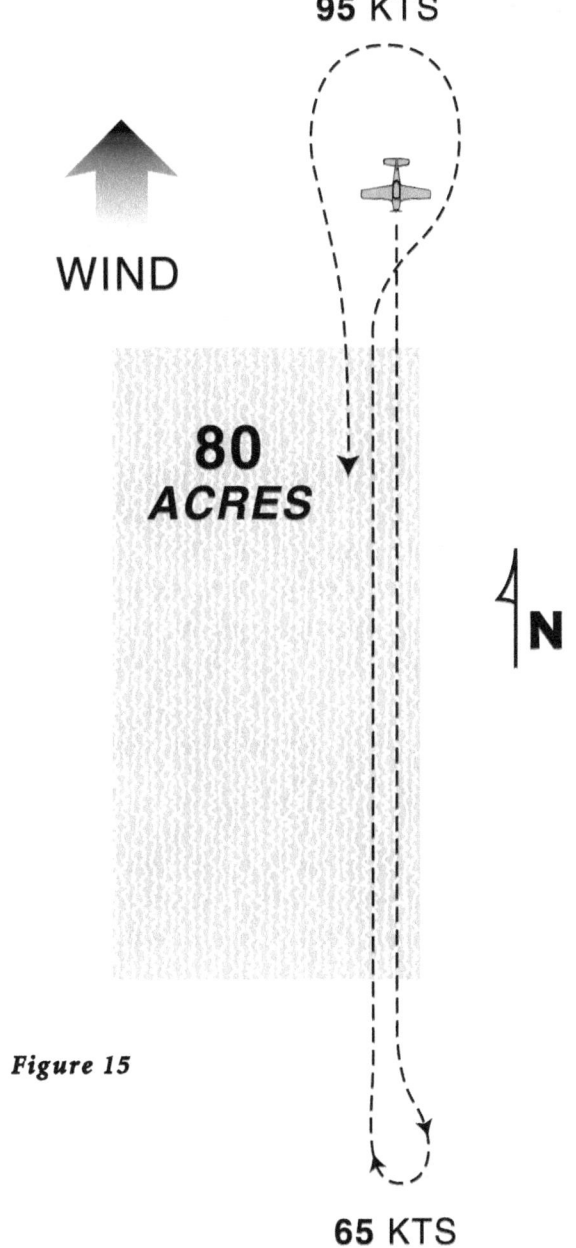

Figure 15

CONTACT FLYING

The problem was that, with the strong south wind, I had to fly a much wider circular return to target on the north end of the field than on the south end of the field. This was because my groundspeed in the north turnaround started at 95 knots (airspeed 80 knots plus 15 knots wind) verses my groundspeed in the south turn of 65 knots (airspeed 80 knots minus 15 knots wind). The result was that I had to fly around at 50' AGL cross country north of the field on each turnaround and re-acquire the field because I had lost sight of the field all thirty seven times when north of the field. I never lost site of the field when south of the field. With airline captains who write downwind turn articles, I beg to differ. The downwind turn does make a difference to some pilots.

The downwind turn has long been a controversial topic for the highflying community. Many times in forty years of flying I have seen the same article about how the wind theoretically has no influence on the aircraft in a turn. It has been established again and again that since the aircraft is part of the air mass moving across the earth, airspeed will not increase while downwind nor decrease while upwind. All this is quite true but not relevant. The relevant safety issue has to do with groundspeed not airspeed. And since groundspeed is a concern in low level ground reference turns and not in high altitude instrument turns, the FAA discounts the whole issue as if it were a myth.

How can an Agency responsible for safety discount the needs of a major group of its customers? We who studied Western Civilization in college learned about trial by oath in the Dark Ages. When they went to court each party to the dispute took the person of highest social rank who would testify in their behalf. The court would then rule for the party having the oath taker of highest social station or rank. Eight centuries later our modern FAA rules there is no problem with downwind turns because Airline Transport Pilots who fly large jets at 40,000' say no problem exists with respect to downwind turns. While there are Agricultural Pilots who hold ATP Certificates and fly jets, they do not fly large multi-engine fan jet aircraft at 40,000'. Therefore the FAA court finds those of us who fly low and actually have real problems with downwind turns trumped by those who fly larger, more expensive jet aircraft, and who fly at a higher altitude. What gets lost is that the problem does exist for low and slow pilots. Our case is thrown out unfounded in light of the rank and social standing of those who know, based on textbook data and irrelevant high altitude experience, that there is no problem with downwind turns. The downwind turn is a real problem for contact pilots.

The real problem of the downwind turn has nothing to do with airspeed. The airspeed stays the same whether upwind or downwind. The problem has to do with high groundspeed while maneuvering aircraft near the ground and obstructions. The spray or pipeline pilot is not concerned with airspeed, which deteriorates and then remains constant throughout level turns or with the airspeed indicator, which records that there is no difference between upwind or downwind portions of the turn. What the low level pilot is concerned with is groundspeed, which increases in the downwind portion of the turn and decreases in the upwind portion.

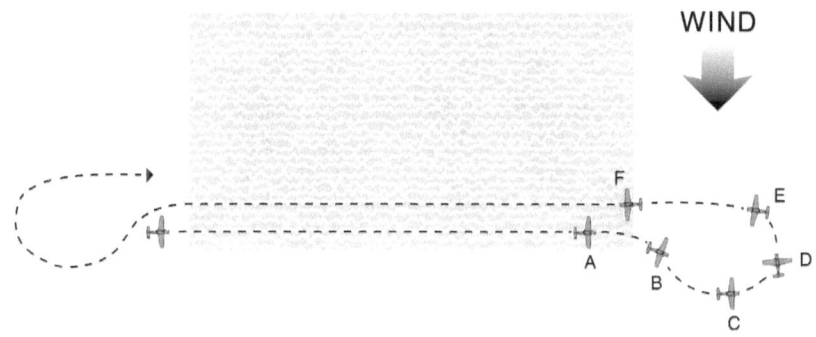

Figure 16

It is this increase in groundspeed that makes the downwind portion of any low level turn difficult. We are not talking about a turn about a point at 1500'. We are talking about an overloaded Pawnee at 50' trying to miss trees, poles, towers, terrain, and tall trucks on the freeway. If the wind is 20 knots and our airspeed is 80 knots, then it is a lot easier to miss these obstacles at a groundspeed of 60 than it is at a groundspeed of 100.

We can do something about this downwind turn problem if we will admit that it exists and that it kills pilots every year. When we have a problem finding an answer to a low level concern and we are unsure about solutions offered by the text, the instructor, the ATP, or even the FAA, we need to walk across to the smelly side of the field and ask the crop duster pilot. *See Figure 16.*

He will tell you that he turns downwind first as he comes out of the field because he has more airspeed at this juncture than at any other time during the turn. Next he will exchange this airspeed for altitude, still during the downwind portion of this turn (B). This puts him up higher during the fast groundspeed portion of the turn. While groundspeed is increasing, airspeed is decreasing as it is being exchanged for altitude (C). This decrease in airspeed decreases the radius of the turn. The slower you go (airspeed) the faster you turn. Now he turns back into the wind (D) at a decreased airspeed (pitch up) and decreased groundspeed (upwind) and allows the nose to fall through naturally. The altitude gained may have been only 50' but at the very time when he is descending rapidly, facing those obstacles, and looking for the correct crop row (E), he is at the slowest groundspeed possible. Finally as he enters the field with nose down and wings level (F) he uses natural gravity (trading altitude for airspeed) to regain zoom reserve in the form of airspeed. He will load the aircraft in the pull up to level off in the field but this is not the load factor that kills pilots who insist (because they were trained that way) on turning level and avoiding the ground. Using the energy management turn rather than the level turn avoids the load factor problem associated with the level turn. Load factors are created by the extra load of turning and pulling up at the same time. This pull up, either to maintain the same altitude or to miss the ground, is what causes the load factor.

This same principle, turn downwind first and save the slow groundspeed of the upwind portion of the turn for the critical return to target (*See Figure 16*), can be applied to other low level operations. This reduced airspeed from pitching up and reduced groundspeed from turning back into the wind makes it possible to line up with crop rows, pipelines, runways or ground targets of any kind. We are concerned with the maintenance of zoom reserve sometimes in altitude and sometimes in airspeed. We are concerned with low groundspeed to best handle obstructions around us. We just have to manage both manmade and natural energy.

During this entire turning, both downwind and upwind, airspeed is of concern only in the maintenance of zoom reserve; that is in maintaining enough buoyancy to prevent mushing into the ground. While much airspeed is desirable in the maintenance of zoom reserve, much groundspeed is not desirable in the maintenance of lateral separation from ground based objects and terrain. The faster the groundspeed the greater is the difficulty in getting properly oriented with the desired crop row, pipeline right of way, or runway. The faster the groundspeed the harder it is to put a wing down between obstacles.

On the other hand, the slower groundspeed of the upwind portion of the energy management turn allows more time to maneuver the aircraft accurately. With a strong headwind our groundspeed could actually be zero. The same headwind that allows us to easily hover-taxi a small aircraft also allows us to safely fly the same aircraft between obstacles with less than ten feet of clearance on each side.

This downwind turn problem exists equally for helicopters and airplanes. Mushing in (settling with power) is mushing in whether your wing is fixed or rotating. Other than in a hover, the helicopter is just an awkward looking aircraft with a big round wing.

In working around objects on the ground we must consider wind direction so that we may utilize our providential gift of slower groundspeed. If the authors of downwind turn articles want to put their butts on the line with their theory about no real difference in upwind or downwind turns, let them make left base to final (*See Figure 17.*) along route A. No fair quoting regulations to escape the problem.

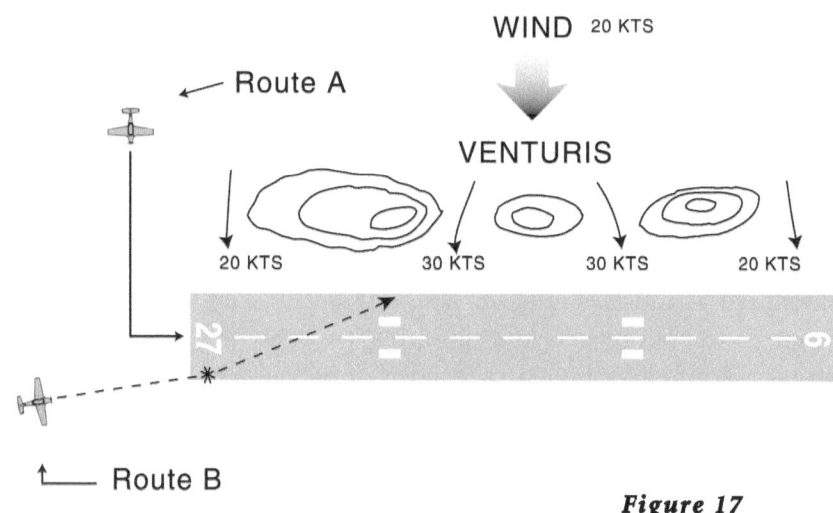

Figure 17

The aircraft on route A will run out of right rudder. The crosswind component is too high. He may not be able to get down safely after going around two or three times. He may not be able to make left traffic on the go around. He may not find fuel close by at an airport with a north south runway. He could however land on the road and fill up at the service station in almost any town in the desert. In small aircraft commercial operations we work with this scenario all the time. I will fly route B using full flaps and land at an angle to lessen the crosswind component.

We can regulate against flying in the West on windy afternoons (most of afternoons are windy in the West). We can regulate against maneuvering flight (the new buzz word for buzzing). But we cannot prevent low level contact flight operations. Every takeoff and landing is a low level contact flight operation. I believe we are not fulfilling our responsibilities to pilots if we use regulations and sanctions in a futile attempt to make maneuvering flight operations go away.

I also believe that Practical Test Standards burden instructors with instrument proficiency numbers that make safe contact flying (looking out the window most of the time) difficult. It is unsafe to force the student to pursue infinite numbers on gauges. He should be flying by reference to the earth. It is unsafe to force the instructor to continuously check numbers. He should be observing things outside the aircraft. The human brain is capable of flying aircraft quite accurately without gauges. I attended Army Flight School with a student who had been a Golden Knight for six years. He used his eyes and brain to determine the proper altitude to open his parachute because observation of his wrist altimeter was less accurate.

The AOPA Air Safety Foundation found that, between 1992 and 2001, 70% of instructional fatalities (# of accidents) occurred during dual flights while only 30% of instructional fatalities occurred during supervised solo flights. Could this be because both pilots spent more time on task with instruments and infinite V speed requirements during dual flights? Instruments have opened a whole new frontier in aviation but they have not come anywhere close to the discernment power of the human brain and our five senses in contact flight. We need for safety sake to get back to the basics of contact flight.

Chapter 8

THE LOW LEVEL FORCED LANDING

In 1991 I flew my biennial flight instructor recertification check ride with an old Marine who flew a Challenger for a bank in Las Cruces, New Mexico. He knew that most of my flying was spraying crops and that I never set the altimeter before takeoff. Very high over the airport he pulled the throttle and announced engine failure. I had a hard time judging the altitude to determine when to stop spiraling. I rolled out high requiring full flaps and full rudder against full opposite aileron to make the last one thousand feet of a mile long runway. He laughed all the way down.

There are differences in perspective and values depending on how we normally fly. If we fly high there are checklists and procedures that must be incorporated to be safe. When we never go above 500' AGL these same checklists and procedures can hurt us. Some of us never fly high. But all of us, the great and the small, fly low at the beginning and ending of every flight.

When spraying the average size field of forty acres, we have less than two seconds on each swath when this field is usable for a forced landing. Those seconds would be at either end of the field halfway through the crop duster turn. If we lose the engine in the field we use the zoom reserve we have stored in airspeed while in ground effect to zoom up high enough to see a nearby field. We immediately turn, very steeply if necessary, to make that field. As we have but four or five seconds until touchdown and obstacles are about, we cannot safely avert our eyes for any reason. Taking our eyes off the landing site just long enough to find a switch or lever could very easily cause us not to make the site or we might hit something. Once oriented toward the chosen site we must immediately begin to slow the aircraft quickly. We will almost always be high and fast. We must be able to find the flaps without looking down.

Please don't use textbook theory to question this, as I have done this very thing several times. Only twice in ten engine failures did I have enough time to even think about anything except the landing that was fast upon me. Both times I did things that hurt more than helped.

This was the case when I ran a Cardinal out of gas on a pipeline in northeast Texas. Three hours and forty minutes into what was planned to be a four hour flight, I was two hundred feet above a crude oil line at point A . *See Figure 18.*

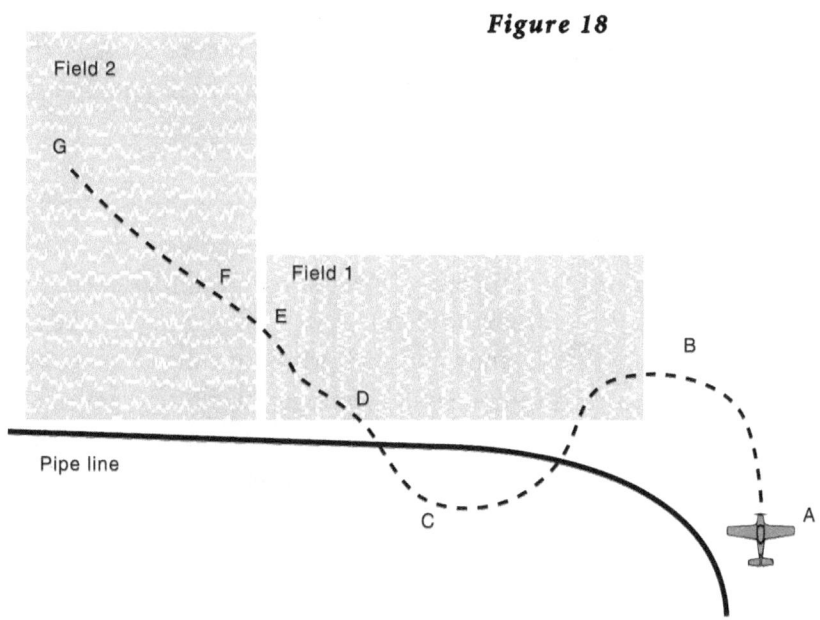

Figure 18

I set up a base leg to field 1, a pasture with a few small mesquite bushes in it. I put the flaps down and, having everything under control with a second or two to spare, I did a very stupid thing. After eighteen years of spraying and training my mind away from such stupid thinking, I tried to make the engine run. I switched the useless boost pump (in a high wing airplane?) on. At point B the engine caught burning the small amount of fuel trapped in the useless boost pump. This little burst of power put me high and fast for field 1. I immediately rolled into a 90 degree left bank. Next I banked hard right with full left rudder to reset for the east end of field 1 and loose altitude. Still too fast at point C I had to push it onto the ground at point D assuming I would knock the nose gear off. Decelerating somewhat I still had enough speed to jump the fence at point E, re-land at point F, and roll to a stop at point G. I didn't use brakes because I knew I had plenty of room once I had jumped the fence.

We might teach all our low time students to always use brakes in a forced landing. If they have made all their prior landings with 1.3 Vso fence speed they will definitely need them. But a well trained contact student need not immediately go to the brakes. Once you have lost a nose gear and got the aircraft going sideways, you become a passenger.

My only other lengthy forced landing was in the late seventies. I left Santa Fe, New Mexico in my Cessna 175 after a drill weekend with the 717th Medical Evacuation Detachment (Helicopter Ambulance). While climbing to cross the continental divide north of Mt. Taylor the engine quit. I was over the Rio Grande and had plenty of time to circle around and set up for the road south of Santo Domingo Pueblo. Because the GO-470 Continental engine in the C-175 is geared I appeared to have partial power with nearly two thousand engine revolutions per minute. However I could count the prop revolutions because the prop was going that slow. At the same time the engine was making horrible grinding and scraping noises and sparks were flying out from under the cowl. I should have shut it down and possibly saved some of the four thousand dollars in parts and labor to rebuild it. Had it been a low level forced landing, I would have been on the ground before the engine had completely beaten itself to death.

How do we prepare contact students for a six second forced landing and high flying students for the last six seconds of a forced landing? What about the engine failure on takeoff? Do we really want to just hit the first obstacle on takeoff in a wings level attitude? I agree with the emphasis on touching down wings level but six seconds is a very long time to sit wings level and do nothing to better our situation. With the zoom reserve either stored in airspeed or altitude anything from 90 degrees left to 90 degrees right is available to a properly trained pilot. We just have to pick a site very close and keep our head out of the cockpit. The turn to the site must be made quickly with bank sufficient to line up as favorably as possible on that site. We will have to lower the nose to prevent load factors. This means low level training in very steep, no load turns (energy management turns) will be necessary.

If a student looses an engine on takeoff (with climb trim), he will have an unnatural kinetic control feel and position. In the takeoff forced landing he needs to immediately pick a near site, turn to it, and kill excess zoom reserve whether in the form of altitude or airspeed. We cannot make an unexpected landing shortly and safely with any excess zoom reserve either in the form of altitude or airspeed. As currently taught, 1.3 Vso is .3 too much zoom reserve in the form of airspeed to land short and safe just as 1000 feet of altitude is too much zoom reserve in the form of altitude.

The airspeed indicator is both dangerous and inaccurate. It is dangerous because we should be looking at the landing site not the airspeed. It is inaccurate because the airspeed indicator does not respond instantaneously and it is but one piece of information. Airspeed is a slow and artificial visual clue. A well trained brain can pick up kinetic, aural, and sensory information in addition to visual. Also the visual apparent rate of closure outside the aircraft is instantaneous and thus superior to slow, artificial airspeed indications on a gauge. The position of the stick can be sensed kinetically and is more accurate than the airspeed indicator. At this juncture we simply do not have time to look at the airspeed indicator anyway.

CONTACT FLYING

The very hard part in this takeoff forced landing, for the highflying aviator, is that he has insufficient time to use a detailed checklist. The resource management sector teaches us to research completely and then act. While this response is useful in the high altitude emergency, it is much too slow in a low level emergency. In the low level or takeoff emergency using the checklist may kill us. What is needed for those who fly at altitude except during takeoff and landing is a simple aural checklist. They could be taught to say, "Below one thousand feet forget checklist, keep head outside the aircraft, turn to miss obstructions, slow down, and level the wings before touchdown." I know I'm asking a lot here but if equal numbers of high and low simulated forced landings are taught the brain can accept differential check lists. Helpfully the new (LSA) Practical Test Standards do not mention a checklist for the simulated forced landing.

Low level emergencies require more than just quick reactions. We must have already accepted our vulnerability prior to the emergency. We must be spring loaded to it. When the engine quits we must have already started for the field. The idea that this cannot be happening never occurs to one who constantly operates in the failure mode. When it happens he turns to the landing site, slows down, levels the wings, and touches down in the time it took to read this.

The down side of this low level mindset is that we occasionally find ourselves in the farmer's field with a workable engine. The up side is that we will be alive and have a real life iteration of a low level forced landing. FAA acceptance of this kind of thinking and training would result in fewer fatalities. Admittedly it would also result in more incidents.

While instructing at AG FLIGHT in Georgia both my student and I forgot to change tanks on a Super Cub during spray training. The engine quit coming out of a spray run. I took the controls, rolled coordinated into a 90 degree right bank, applied full left rudder to slip while still banking right, pulled full flaps and was on short final to the adjoining field by the time he had changed tanks. The engine caught but I pulled the throttle and landed anyway. You just don't have time to consider correcting the problem on a low level forced landing. I could have reasonably assumed the engine had quit because of fuel starvation but what if I had been wrong?

We cannot teach low level forced landings if the minimum safe altitude is an arbitrarily set 1500' AGL. This is a reasonable floor for high altitude forced landings. But our failure to teach and practice low level forced landings could contribute to a fatalistic attitude about them.

We all leave the ground and return to it on every flight. Engine failures often follow power changes with takeoff power contributing its share. It is dangerous to assume that spending most of every flight at high altitude will statistically preclude any need for low altitude forced landing training. In the mountains the assumption that we have the power to stay high above the ground even during the enroute phase is false and dangerous.

The engine failure on takeoff is not a desperate situation. It is no different than the one many spray pilots have experienced at the end of a spray run. We simply pitch up wings level trading speed for altitude until a nearby landing site is selected. We then turn immediately with whatever bank is necessary to make that site. We allow the nose to fall through preventing load factor. We decelerate and apply full flaps (if available) to make the very beginning of the site using the brisk walk apparent rate of closure.

Decelerating rapidly to make the site is not a problem, if we have practiced landing on the numbers every time using the apparent rate of closure approach technique. The angle of decent will be somewhat greater because of the drag of a windmilling prop. Having done this ten times and never ending up short of the field, I can say our extra caution to save altitude more than compensates for the windmilling prop but the rate of descent is shocking.

CONTACT FLYING

In 1991 I leased a Piper Pawnee from an AG operator in Missouri to spray cotton, chili, and vegetables in the Mesilla Valley south of Las Cruces, New Mexico. This operator worked on a handshake lease agreement that was easy on the pocketbook but his aircraft were not the best. This engine was only 50 hours out of major but began heating as the season progressed. The temperature was rising to eighty degrees before ten by mid June and I had had to make several precautionary landings with the engine cutting out. Flushing the oil cooler and installing new plugs didn't help.

While I was applying Trigger on cotton to open all the bowls the engine quit on a pull up. I turned to an open area in the next cotton field. There are areas completely void of crop after laser leveling. I hoped to slow down in a cotton free area before hitting the tall cotton. I slowed up too much thinking I was at zero thrust with a poorly running engine. I began to fall through and added some power. Nothing! I pushed over into ground effect making the bare spot but landed hard. The tires mired in the deep, soft crop rows and she went over on her back. The Pawnee canopy went down between two rows leaving loose dirt outside both side windows.

When on our back we tend to get disoriented. Hanging in my harness I saw liquid draining rapidly from what I thought was the tail of the airplane. Since I was looking the opposite way from how I had just been going, I thought I was looking at the tail. I wondered how my load could be pouring out the side loader located on the fuselage behind the pilot. What I was actually seeing was gas pouring out of the broken nose tank.

Even disoriented I had plenty of adrenaline to help me unstrap, fall to the top of the cockpit and butt through the Plexiglas window with my helmet. I dug out like a dog going after a rabbit.

Walking away from the aircraft I saw a lime green fire truck speeding down a dike road toward the edge of the field. I ran over to stop them. After much discussion I finally persuaded three eager firemen and one eager firewoman not to trample the farmer's expensive pima cotton crop. The Pawnee did not burn.

Most forced landings follow one to three stupid acts. In corporate or airline flying it makes perfect sense to work on eliminating all stupid acts. In combat, crop dusting, or pipeline flying, on the other hand, just being there is a great risk to begin with. We best learn to deal with the calculated risk. Should we fail, our calculated risks will be called poor judgment. In the Cardinal incident I was cited with "careless and reckless" flying. By the way, always file the NASA incident report. You will thereby invoke immunity unless the FAA can find clear evidence of a violation. According to Murphy's Law and King Solomon's Proverbs, humans are by their very nature stupid.

While spraying vegetables in Fabens, Texas I had a spray pump go out on my Callair spray plane. It had been over one hundred degrees for forty days and over 110 for fourteen days (summer of 94). My loader and I spent the entire day trying to find a used pump that would work. This was the first stupid act.

We kept putting my good plastic pump fan on used Transland pumps until finally my loader asked, "Couldn't this fan (he was holding one of the beat up fans off one of the used pumps) be different? I said, "Transland is a reputable company. They wouldn't do that to me." Well that used pump had a Russian pump fan (rotates counterclockwise). My plastic fan needed to rotate clockwise to work. We put the Russian fan on the pump and I went out to calibrate for the third time that day. But before going I committed the second stupid act. I changed my routine.

Because I am a naturally disorganized person and sometimes Murphy's best friend, I never attempt fuel management when it can be avoided.

CONTACT FLYING

High volume vegetable work is 15 minutes per load in a Callair. Therefore I carried ten gallons working fuel in the left tank and five gallons reserve in the right tank. This had been working well for me; I hadn't had to switch tanks all season.

Before going out to check this Russian pump and calibrate I asked my loader to put ten gallons in the left tank. He asked if I couldn't just use the reserve in the right tank assuring me he would set both tanks up normally before I went to the field. He did as stated but I never changed back to the left tank. This was the third stupid act.

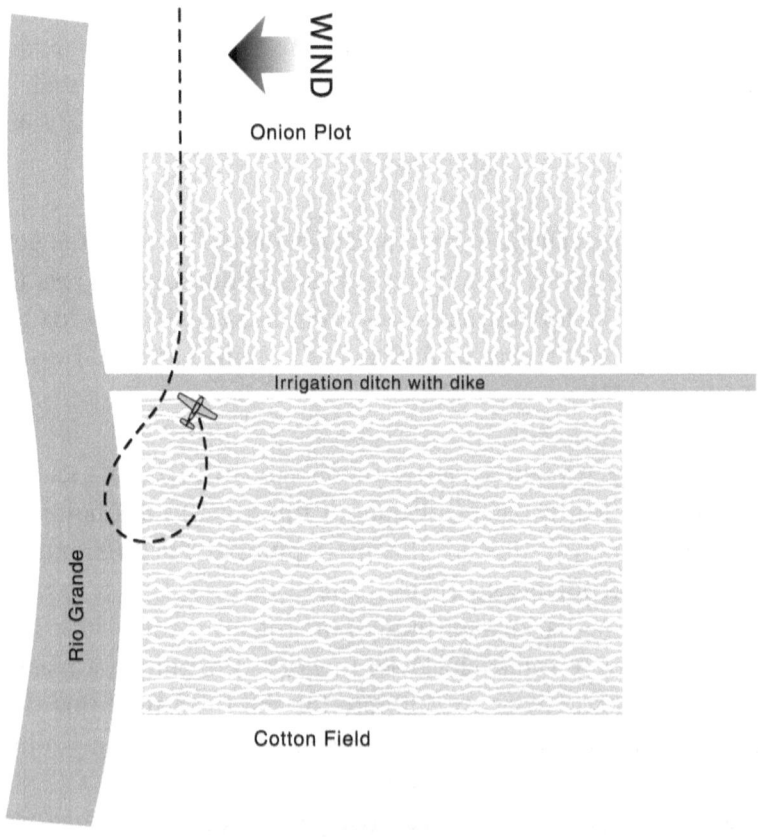

Figure 19

This forced landing was much different than the six previous ones. With the adrenaline boost, one is usually fully alert. This time I was fully relaxed, even euphoric. I assume I was high on the residual Parathion from the used pumps. The pyrethroids I was spraying were not that toxic. What I saw drift leisurely before my eyes (*See Figure 19*) was my inability to make the onion field I was spraying. There was a deep, wide irrigation canal with a levee intervening. I saw I could make the cotton field I was turning over and began a turn to align with the rows. I saw I could not make the turn to align and I thought I had leveled the wings prior to touchdown.

What really happened after the engine quit, as the border patrol agent on the levee of the Rio Grande later told me, was that I banked into the ground and cart wheeled onto the levee that bordered the onion field. The right wing spar was broken in two places (cart wheeled twice off right wing and nose) and the engine was hanging onto the firewall by cables only. The Lord truly cares for fools because I walked away from this crash like all the others except Vietnam.

Thinking about that poor Callair sitting upright on that canal levee with poison running out the broken right boom and gas running out where the gascolator had been brings to mind an interesting story. The next day found me sitting in the local diner discussing cleanup of the canal with the local EPA Agent. My broken boom had leaked into the concrete canal. It was 50' across the top, 20' across the bottom, and had 4' of water in the bottom. One quarter mile up the canal was the gate letting water in from the Rio Grande. One mile down the canal was the next gate. Another mile down the canal reentered the Rio Grande. The EPA Agent was just telling me I would have to drain the canal completely between the gates when my grower walked in. This was of course going to bankrupt me, a one plane one pilot operator. The grower, whose onions I had been spraying, walked up to our table and asked if we were talking about the spill? Before the EPA Agent could answer he continued, "Well you don't

CONTACT FLYING

have to worry about that because I opened the gate and let that stuff go on down the river." The Agent slumped and said nothing as the grower sat down beside me. Finally he got up and walked out saying he might contact me later. I got a $1600 bill for the cleanup on the ten gallons of gas I lost out of the left tank but as the Almighty had saved my life this grower had saved my business.

Update: I just had engine failure number ten this morning in an Army Stearman Kadet looking homebuilt with a Subaru engine. The owner wanted me to test fly it before checking him out in it. I like manufactured airplanes because you can depend on the way they fly. This homebuilt flew well but car engines are not as dependable as airplane engines. This one had two carburetors on it and a single ignition. For some reason it ran rough and quit. I landed in the farmer's field, knocking the right gear off on his fence on the way in. The plane was very stable all the way down and slid well on the lower right wing and left gear in the sloping field of wet orchard grass. The slope worked out to put my good left gear downhill from my no right gear. Except for breaking the right gear, I did only cosmetic damage to the airplane.

I was flying Midwest pipelines with a C-172 that kept cutting out on me. I had made several precautionary landings on roads. We had checked plugs, mags, and carburetor with no joy. I found I could keep it running by keeping the mixture full rich. We normally lean Lycoming engines, even at low density altitude, because they run better that way. After checking my fuel burn the boss said to run it lean. I said I would but if I had to put it on the road again I would leave it there.

Shortly after this I was at Snyder, Texas for my one hundred hour inspection. I was helping the mechanic put a wing tank in another C-172 that he needed to finish first. The tank was dusty all over and was not all taped up and steam cleaned as if it had come back from the welding shop. My foreman came into the shop and told me that this airplane, seven eight Juliet, was now my airplane. We finished the tank installation and I flew pipelines to Dallas, Remain Over Night at Jackson, St. Louis, and had stopped for lunch at Lebanon, Missouri on Shell's Ozark Pipeline. After lunch the engine quit on takeoff shortly after liftoff. I pulled power, added full flaps, and landed on the runway remaining. We found mouse hair in the thumb screen to the carburetor bowl. We also found a dead and mummified mouse in the tank.

If I had succumbed to the peer pressure to make normal takeoffs unless overloaded or in the mountains, I would have been two hundred feet up when the engine quit. I always get the aircraft off the surface as soon as possible but stay in ground effect until I achieve what is called zoom reserve in the form of airspeed. The last time I made a "normal" takeoff was on my Instrument Instructor Practical Test in 1976. I was very happy to be able to remain in ground effect and hold off until touchdown on this forced landing. I was very happy not to have to wonder if I would be able to get it down on the runway remaining from 200.' I have had forced landings to the jungle, roads, pastures, cotton fields, and the levee on the Rio Grande. I had more than six seconds to worry about it on only two of them.

CONTACT FLYING

Chapter 9

LOW POWER MOUNTAIN FLYING

Map 1

In the seventies I was flying east of the Sangre De Cristo mountains in a Tri-Pacer from Las Vegas, New Mexico to Denver. Standing lenticular clouds were lined up along the high peaks indicating high winds and turbulence in strong updrafts and downdrafts. At 13,500' MSL I thought I was well above the declination line north of Las Vegas VOR. *See Map 1.* I hit a lee side downdraft so violent that I was covered with dust from the floor. In less than a minute I was at 6,500' MSL. A seven thousand feet per minute downdraft will ring your bell and cause you to make peace with your Lord. I stopped at Springer Airport to clean up a bit and repent. "Be still; and know that I am the Lord." Psalm 46:10

Oxygen was installed on our Hueys in the Army's Mountain Flying School at Ft. Carson in Colorado Springs because of a wave incident. An instructor with two students and a crew chief hit a wave at 8,500' MSL, bottomed the collective (in calm air the UH-1 will descend at 3,000 feet per minute at flat pitch), and ascended uncontrollably to 17,500' MSL. There is vast natural energy in the mountains and it is not manmade. Either the 150 horsepower Lycoming O-320 engine in a Tri-Pacer or the 1300 shaft horsepower Lycoming T53-L13 engine in a Huey pales in comparison to this tremendous natural power.

We must have faith in the vast energy that God placed in nature and not be so dependent on large engines to find our way through the mountains. Compare the one hundred feet per minute climb ability of our 150 hp engine at ceiling to the three thousand feet per minute up and down drafts any give afternoon in the mountains. The desert weatherman gets his daily high temperature estimate by recording the temperature at BMNT (Before Morning Nautical Twilight) then adding 30 degrees in winter or 40 degrees in summer with adjustments for cloudy days. Albuquerque International Airport has .8 weather days per year. This is vast energy and it is dependable. The Rocky Mountains rise into a jet stream that typically produces winds aloft in excess of two hundred miles per hour. This is vast energy and it is dependable. Our little engine produces one hundred feet per minute of climb at 10,000' MSL on a standard day. This is not vast energy and it is not dependable.

Winds, gusts, thermals, standing lenticular clouds, high rugged terrain with deep canyons, and a forty degree temperature spread between first light and heat of day will cause the chart calculators to recommend "no fly." Not to worry. This instability is the source of greater energy than that provided by Lycoming or Continental. By artfully harnessing this natural energy, we can regularly climb at one to three thousand feet per minute and safely make sixty to ninety degree bank turns in mountainous terrain. But we cannot capitalize on these energy sources unless we put our faith in those natural laws that control this natural creation and sharpen our feel for what is and what is not available to us. Most of all we have to fight the habit of relying on engine power.

Nature has provided five sources of energy in the mountains and high desert. They are ground effect, thermal lift, gravity, hydraulic lift, and mountain wave. Ground effect will help us get off short fields, cool our engine in the desert heat, climb alluvial slopes, and safely cross mesas and wide mountain passes. Thermals are always a source of altitude if we spend more time in updrafts than in downdrafts. Gravity is a fast source of airspeed in narrow valleys where lateral space is limited but vertical space is abundant. Hydraulic lift (the Venturi effect of air passing over uneven terrain) will make it possible to cross the high mountain passes. Mountain waves (sympathetic hydraulic lift) allow high mountain pass westerly crossings on really windy afternoons.

Two of these natural energy sources are fixed. The other three are variable. Gravity and ground effect are fixed. They are always capable of producing zoom reserve in the form of airspeed. Because gravity and ground effect are always present, we can bank on them. At every moment during flight, but especially during takeoff, we consider the direction to lower terrain and any obstructions that might limit our taking full advantage of gravity and ground effect. Aircraft go faster downhill than in ascending flight. Aircraft go faster in ground effect than at altitude. Thus we will turn to lower terrain in emergencies. We will avoid obstructions laterally even when this requires a steep turn at altitude or a cross controlled wings level rudder turn when in ground effect.

Fifteen degrees of yawing turn can be accomplished, in ground effect, by fighting the wing (or tip-path-plane) level while pushing the nose around with rudder pedals (or anti-torque pedals). Steeper turns can be effected safely by pulling up to clear the wing or tip path plane and then allowing the nose to fall through naturally to go back to ground effect after the energy management turn. Whatever the method, we avoid obstructions laterally until we have achieved sufficient zoom reserve in the form of airspeed to safely climb out above the obstructions. And when we do attempt the climb out, we keep valleys and obstruction free routes in mind for emergencies.

Thermal lift, hydraulic lift, and mountain wave lift energies are variable sources of much more powerful energy than that produced by our engine. Because they are variable, we must be artistic in our pursuit of them. We must learn to seek these natural highs to cool our engine while increasing our performance.

The FAA teaches avoidance of the natural power available to the artistically flexible. Instead they promote financial (bigger engine) solutions to mountain flying problems. Given enough bucks and a big enough engine, you may think you have "reached out and touched the face of God" as the poet suggests. It would be wise, however, to know the strength of the natural elements. "Seek ye first the kingdom of God," and 3,000 fpm rates of climb (zoom reserve) will be added to you.

The techniques involved in takeoff are the same whatever the location or aircraft type (airplane or helicopter) but altitude, like load, will test both pilot and engine. Normally aspirated engines develop significantly less than full rated power at high density altitude. Your POH will give you the percentages but it will not prepare you for that sick, sinking feeling that comes with altitude. Gravity and ground effect must be kept in mind during mountain takeoffs. We must know the direction to lower terrain (a gravity consideration) and we must have knowledge of any obstructions that might hinder our getting to lower terrain in ground

CONTACT FLYING

effect if necessary (a ground effect consideration). Most mountain and high desert airports are near a river or wet season drainage. We need to know which way the water drains to utilize the tremendous force of gravity.

Below 4,000' MSL ground effect may be critical only on short fields for building zoom reserve in the form of airspeed to clear obstructions. In the more populated lower areas of the world obstructions abound. Above 4,000' MSL ground effect is necessary on nearly every takeoff of a small aircraft in the summer. Above 4,000' MSL there is less population created obstructions and fewer trees. Developing cruise speed may require flying downhill in ground effect.

Any time we get into power problems or severe downdrafts we need to immediately turn to lower terrain. Knowing the direction of drainage is critical. Downstream is usually obvious in the West but not as easy to judge in north Arkansas, the Adirondacks, or the Appalachians. I cannot overemphasize the need to fully utilize ground effect and gravity in high country. These natural forces can prevent the loss of Cardinals and seven year old pilots, even if the plane is overloaded. Not only do we need ground effect to clear obstacles off the end of the runway (they seldom exist at high fields), but we may need to stay in ground effect until we are well away from the airport while missing obstacles laterally in our search for lower ground. Once zoom reserve in the form of airspeed is established we can turn uphill and begin the search for good air.

Thermal energy is also available in the Midwest. In the West however, the daily temperature spread is larger and more dependable. Using thermal energy, rather than our air starved engine, simply requires flying slow in updrafts and fast in downdrafts for a net gain in altitude and airspeed over time and distance. To help ourselves learn to kinetically, aurally, and visually detect updrafts and downdrafts, we need to observe our tachometer on fixed pitch propeller aircraft. If we use pitch control to keep the RPM at cruise setting (don't touch the throttle) we will fly slow in rising air and fast in descending air. We can also use the vertical speed indicator (VSI) as a crutch until we learn how to feel updrafts and downdrafts. When the VSI indicates we are going up at more than 500 feet per minute while we are in a level attitude we should pitch up. This slows the aircraft so that we stay in the updraft longer than if we had not pitched up. The VSI will now indicate a very stable and high rate of climb. When the VSI begins to wiggle, even though it still indicates significant positive vertical speed, we need to level the aircraft. When the VSI indicates a descent of more than 500 feet per minute while we are in a level pitch attitude, we should pitch down. This speeds the aircraft so that we stay in the downdraft less time than if we had not pitched down. Don't worry about altitude. The downdraft physically cannot go into the ground. The stronger the downdraft the higher it will compress on the desert floor.

Hydraulic lift is available anywhere strong winds pass over rising terrain. Air over major terrain features acts like water over rocks in a river. *See Figure 20.* As a part of the air mass we will be swept up and over the terrain feature like a cork in the water would be swept over the rocks.

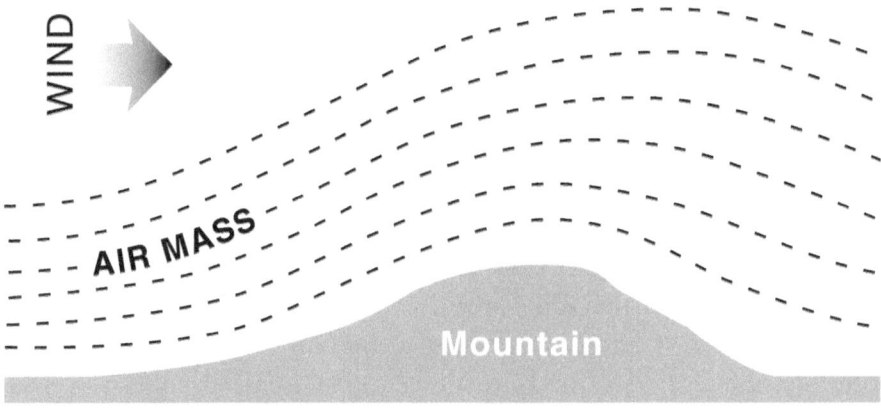

Figure 20

Mountain wave results from a sympathetic action like the marching of troops across a bridge to a cadence. This repetitive, sympathetic action (hupt-two-tharee-four) will destroy the bridge and injure our troops. It happens in nature like this. *See Figure 20a.* First a small peak downwind just far enough receives tremendous plunging air off a higher peak upwind. This air ascends the small peak with a whip action. If at the right speed and distance to accelerate with this second lift, sympathetic wave action can come into play. The force of the updraft over the downwind peak is stronger than the force of the updraft over the higher peak. A third peak downwind at just the right distance (frequency) will continue increasing the speed of the air over it.

Figure 20a

Lenticular clouds can line up in a chain downwind from high peaks. Gliders use these chains of lenticular clouds (multiple waves) to fly to military jet altitudes in the mountains and desert.

Oxygen was installed on our Hueys in the Army's Mountain Flying School at Ft. Carson near Colorado Springs because of a wave incident. An instructor with two students and a crew chief hit a wave at 8,500' MSL, bottomed the collective to try to descend (in calm air the UH-1 will descend at 3,000 feet per minute at flat pitch), but ascended uncontrollably to 17,500' MSL.

The landing problems peculiar to the West are mesa top airports, one way strips, and strong crosswinds. Mesa airports like Farmington, Sedona, and Las Alamos require high glide slope angles to insure not going below the declination line on short final. Anytime there are headwinds on final to a mesa there will be strong downdrafts on short final as the wind plunges off the mesa. *See Figure 21.* We need to use full flaps and slip as necessary to get down quickly once beyond the drop off of the mesa. This is especially true at places like Las Alamos where we also have a one way field. Altitude is airspeed and airspeed is altitude. We use only what we need of both because we have only one shot at it. We use the apparent rate of closure approach even with the steeper glide slope because there is no go around.

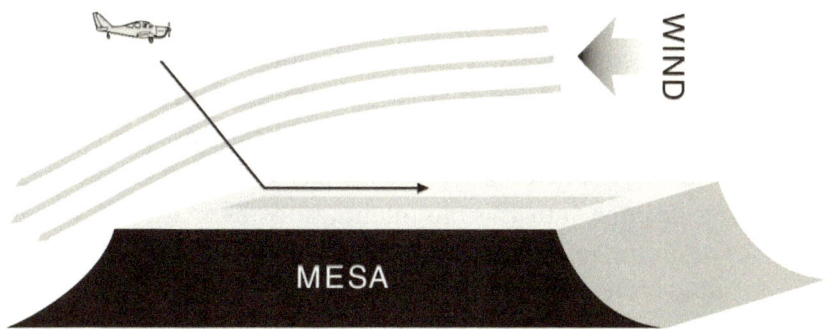

Severe hydraulic downdraft when afternoon winds are high.

Figure 21

One way strips require the apparent rate of closure technique. We must make the field without needing to go around because, again, there is no go around. Teaching that the go around always demonstrates good judgment just doesn't always demonstrate good judgment. If the objective of the FAA is safety, then rote judgment is not sufficient. If the objective is sanctions, then rote judgment is easy to adjudicate.

CONTACT FLYING

We simply use full flaps, a brisk walk apparent rate of closure, the shallowest glide slope possible, and put in on the first 1000' of runway between the downwind corner of the runway and the upwind big airplane touchdown zone marking.

How can we leave Gallup, New Mexico (See Albuquerque Sectional) early on a summer morning in our 172 and get to Albuquerque fifteen minutes ahead of our buddy, Sam, in his 172, burn fewer gallons per hour, and run a cooler oil temperature? Sam studies all available FAA sanctioned information sources to determine best altitude based on climb winds, winds aloft, temperature, best TAS altitude, hemispheric rule, etc. All these numbers erroneously presume Sam has power available to climb to any altitude he chooses. Concluding that the longest time spent at that altitude (especially going east with prevailing westerly winds) will produce the best groundspeed, Sam takes Runway 27, rotates, and pitches up as if his engine will take him at book rate of climb to that altitude. His engine will not accomplish this rapidly at 200' per minute, however it will pull him away from three sources of free energy: ground effect, the heating air mass moving upslope into the mountains, and thermals. By climbing quickly out of ground effect Sam has made getting over the huge highline north of Gallup VOR dicey. Unless it is already hot enough to produce good thermals and he pulls up in updrafts and pushes down in downdrafts, Sam will heat his engine in a very slow book perfect climb to altitude. Later in the day, as the desert heats the cool morning air, thermals will work well.

We too have taken off west from Gallup because of the prevailing westerly wind and down sloping valley. We have set the mixture prior to takeoff to gain three hundred engine revolutions per minute. But rather than trying to go up using book best rate of climb, we stay in ground effect out over the empty fields west of the airport until zoom reserve is achieved. With southwesterly winds we climb the mesa north of I-40 using hydraulic terrain lift, or with northwesterly winds we do the same over the hills south of Gallup VOR conserving zoom in either case. By staying close to the ridge north or hills south, we give ourselves the entire valley for the turn around to go east. By using thermal and hydraulic energy to store zoom reserve in altitude while in updrafts and pushing over storing zoom reserve in airspeed while in downdrafts, and by using ground effect when terrain permits, we take advantage of all natural energy available and run our engine cooler than Sam's.

In southwest winds we ride the mesa northeast of Gallup to Mount Taylor. If the winds are from the northwest we ride the ridges southeast of Gallup to Grants then jump across the valley to ride the northwest slope up Mount Taylor. Thus we arrive at Mount Taylor at the same ten thousand feet MSL as Sam without sacrificing groundspeed or heating our engine. We both will trade this altitude for airspeed the remaining fifty miles to Albuquerque, but we should arrive over the Mount Taylor fifteen minutes ahead of him.

Going back to Gallup in the hot afternoon with winds blowing thirty and gusting to forty five at West Mesa Airport, even worse up high, we may have luck thermaling up, ridge riding, or we may catch a good mountain wave east of Mount Taylor. Rigid application of book numbers, however, will keep us on the ground in Albuquerque. We don't have enough fuel for an engine climb to 12,000 into 60 mph headwinds. Flying west after noon in the Rockies is an art. Science will tell you it cannot be done without turbocharged piston or jet engines. Art and experience in cooperation with natural forces will prove that it can be done if we keep our priorities straight.

In order to get west in the afternoon we must launch with great hope and feel for the good air, but we must also be prepared to turn back when necessary. If we thermal up or catch a wave we may end up above the declination line and cross the pass at a comfortable 500' AGL. We may ridge ride or take a long dive off a downwind wave. We have to feel for good air constantly aware of the direction to descending terrain. We must be vigilant of impeding obstructions that might cut us off from

CONTACT FLYING

descending terrain. When it becomes necessary to turn to lower terrain, we will go immediately and aggressively. We fly 50' laterally to the rocks on one side of the valley or canyon, leaving ourselves the whole mid-valley to turn to if we see we can't make the pass.

The easiest but possibly the slowest ride home involves thermaling up while holding a westerly course. Thermaling in a powered airplane or helicopter is not like thermaling in a sailplane. We don't have the endurance to zero our groundspeed while circling in one thermal. Instead we attempt to fly slowly in updrafts and fast in downdrafts while maintaining a constant course. Until we develop a feel for ups and downs more sensitive than our vertical speed indicator (VSI) and tachometer we use our VSI and tachometer. When we feel a sharp bump we know we have hit a shear. If the VSI goes up say to 500 fpm and our RPM increases indicating an updraft, we pitch up to Vy. The better the air the longer we want to stay in it and therefore the slower we want to go. When we pitch up the VSI will more than double and then stabilize. We maintain this pitch up attitude until the needle begins to wiggle. We keep the RPM stable with elevator. Even though the VSI continues to show say 2000 FPM up, when it wiggles and the RPM decreases we have lost the updraft and a downdraft will follow about ninety percent of the time. When the needle wiggles and the RPM decreases indicating a downdraft, we level the aircraft. If this leveling results in a 500 FPM down indication, we pitch down to a power descent attitude. When we level the aircraft, if the VSI indicates a rapid descent and RPM falls well below cruise, we pitch down to maximum structural speed attitude. We keep the RPM as stable as possible with elevator (throttle is full). The worse the air the shorter the time we want to spend in it.

And here is where a tough but necessary technique comes into play. A really strong downdraft may send us to the desert floor but it cannot push us into the ground. It will compress against the desert floor not taking us below 50.' Crop dusters know the air doesn't go into the ground because when they hit gusts they are always thrown sideways but never down.

This is the compression of the downdraft fanning out horizontally upon contact with the ground. We will feel a really sharp bump at the bottom of the downdraft where we may level the aircraft or push over into ground effect. The RPM will increase. We need not be discouraged about being taken to the deck. Airspeed is altitude and altitude is airspeed. In the mountains we take what nature gives us. Also we are flying into rising terrain so we are not losing that much MSL altitude.

Working the elevator to keep the RPM stable will cause us to go slow in updrafts and fast in downdrafts. When an updraft increases the RPM, we pull up to maintain the same full power RPM. We must check cruise RPM at full power to obtain a baseline. This pull up will slow our airspeed in the updraft. We control RPM with elevator, as the throttle is at full. When we hit a downdraft the RPM will fall off. We push over to maintain the same cruise RPM. Airspeed will increase in the downdraft because we have pushed over. Going slow in updrafts and fast in downdrafts will give us a net gain in altitude without any net loss in airspeed, a gift from nature. Remember we stay at full throttle in the mountains and high desert. However, we never attain 100% power with normally aspirated engines.

Question: won't this pushing over in downdrafts cause us to fly into the ground because we are diving in downdrafts? According to Sir Isaac Newton, what will that descending shaft of air do when it impacts the earth? Will there not be an equal and opposite reaction to that impact? The greater the vertical velocities of the downdraft the higher above ground it will bottom out, spread laterally, and even rise.

We will have a net altitude gain over time and distance so that by the time we reach the high mesa north of Mount Taylor or the ridge to the south we should be well above the declination line. By spending more time in updrafts than in downdrafts we are lifted in our straight line course. When thermaling in this manner in rough air we are usually taken up so high so fast that we spend a long time in tremendously high headwinds.

CONTACT FLYING

Headwinds of sixty, not uncommon up high, will cause a fifty percent reduction in the groundspeed of a 172 or a Cherokee. You have to watch your gas when traveling west at high altitude in the West where the average distance between full service airports is one hundred fifty miles. And obtaining gas requires giving up hard earned altitude.

This technique will work over low flatland as well. Have you caught yourself adjusting the throttle too often in rough air? Throttle adjustments are unnecessary and counter productive. If we pull up to control RPM increases and push over to control RPM decreases, we will have a net altitude gain with no loss in groundspeed.

Map 2

A faster but less certain way to travel west in the afternoon is to fly upslope in ground effect. When we get off the desert floor and into the rougher terrain, we look for any ridges or mesas that run northwest if the wind is from the southwest or for mesas that run southwest if the wind is from the northwest. We can ride up hydraulically but our AGL altitude can never be too high or we will miss the good air and climb into the descending lee winds off the higher ridges on farther up. North south ridges or mesas are of no long term help because you cannot make any groundspeed toward Gallup flying north or south.

I can better illustrate the westerly hydraulic attack in La Manga and Cumbres passes between Alamosa, Colorado and Chama, New Mexico. *See Map 2.* Flying west with winds from 225 at 40 gusting to 60 up high, we would want to follow the left fork of the Conejas River to the second fork in that river north of La Manga Pass. This should give us hydraulic lift on the long northwest ridge north of the river. We should be lifted to over ten thousand and be able to cross La Manga Pass.

The whole way up we must ride this ridge as laterally close to the rocks as possible. If we fly over the river in the middle of the valley we give ourselves no out in case of severe downdrafts. We would hit the canyon wall on either side of the river in a FAA sanctioned level turn at a shallow bank.

The test of nerves will come when we must turn southwest toward La Manga Pass at the third fork up the Conejos river. Here the valley is much tougher but workable. We just ride the narrow ridge that juts north from east of the pass. If we can't find good air there we have two other options. We could ride the original ridge on further northwest past the camp (marked camp on *Map 2*) to the windward side of the 11,201' peak and then reverse course back to La Manga Pass. Or we could go back down the main valley to Canon and try going direct to Los Pinos, Colorado on the south side of the 10,746' mountain over the narrow gauge railroad.

CONTACT FLYING

The high mountain pass turn around involves the same technique as the crop duster turnaround or gunship return to target except that at high altitude it takes little or no pitch up to decelerate sufficiently to turn. A level, shallow banked turn will not work inside a tight canyon.

The valley floor is 8500' northwest of the camp (marked "camp" on *Map 2*). The mountain west of the camp is 12,000' and the mountain northwest of camp is 11,201.' That provides as much as 3500' vertical space under our left wing. We cannot make a fifteen degree banked turn to the left without hitting the south canyon wall. What we must do, if we hit severe downdrafts, is immediately roll into a left turn of sufficient bank to miss the south canyon wall. At the same time we allow the nose to fall through naturally (this is the energy management turn) to maintain the same angle of attack. This may require as much as 90 degrees of bank. In Mr. Langewiesche's chapter on "What the aircraft wants to do," we find that there is no load factor. The aircraft is well designed to maintain a constant relative wind. It will not load itself in a turn of any degree. It will simply pitch down sufficiently to maintain the same relative wind. Only the pilot can produce a load by pulling back on the stick. Solution: we don't pull back on the stick until the turn is completed and the wings are level. We have 3500 vertical feet to accomplish this in. After the 180 degree course reversal at 60-90 degrees of bank, we first level the wings and then level the nose. This leveling of the nose creates a load but not the load factor in the POH that would create an accelerated stall. We must be sure to level the wings before pulling out of the dive. Trying to pull out before leveling the wing could create the graveyard spiral.

Unless we are already high enough to make the pass, the use of a level turn with minimal bank will send us into the canyon wall on the south side of the Rio Conejas. Without training in energy management turns, we are in deep trouble here.

Sanctioned curricula used to teach the lazy eight in a natural, artistic way (the energy management turn) that would have been a useful solution to this problem. We were to use 45-60 degrees of bank allowing the nose to fall through naturally without creating load factors. This turn back in deep, narrow, mountain passes uses the very same technique. The same is true of the crop duster turn and the gunship return to target.

Everything in flying is dynamic and variable. We must use only the amount of pitch up that will slow us sufficiently to make the turn. This will conserve zoom reserve either in altitude or airspeed as available. We must remember that our nose is already higher to maintain level flight at altitude. We must also consider the vertical space available. In desert mountain passes the valley may not be as deep but the lateral space will usually be greater so we must take what nature gives us to work with. We must not rely on the engines of man and rigid training parameters.

A fatal Cherokee 180 high altitude canyon turnaround accident, NTSB DEN99FA157, revealed training problems and misconceptions in mountain flying. According to the NTSB Final Narrative, "An FBI pilot, along with two other special agents in separate airplanes, was receiving agency-sponsored mountain flying instruction. His instructor announced that they were going to practice box canyon turnarounds. Shortly thereafter, white smoke was seen coming from the side of the mountain. Evidence indicates the aircraft struck tree tops while in a steep left descending turn in what was described as an extremely tight box canyon."

Additional Data/Information in the NTSB Preliminary Narrative states, "The two surviving instructors were asked to describe the egress procedures that are taught to students for turning around in box canyons. The first method involved a descending 60 degree bank, with about 10 degrees of flaps deployed. The second method was basically a chandelle, i.e. a 180 degree steep climbing turn. The third method was the 'wing over,' a crop duster maneuver used to reverse direction in a short distance. It involves pulling the nose of the airplane up to a steep angle then, just before the stall, applying full rudder to reverse direction. The FBI interviewed one of its special agents who had flown with the accident instructor in N4366D during the previous course. He said the instructor asked him to perform a box canyon turnaround, and talked him through the first method described above. They dove out of the canyon approximately 100 to 200 feet above the ground following the sloping terrain away form the mountain. The instructor then talked him through the third method described above. He said the nose was pitched up 20 degrees to diminish airspeed, then left aileron and right rudder were used to reverse the direction. The maneuver failed and confused him. He asked the instructor to demonstrate the maneuver. Although the instructor was successful in performing the maneuver, the student was still confused because the control inputs seemed to invite a cross-control stall."

The first method most closely resembles the energy management turn, but the amount of bank should be that necessary to miss the obstruction. We bank sufficiently to miss the canyon wall even if the degree of bank needed is greater than 90. "Descending 60 degree bank" poorly describes the amount of pitch down as well. The amount of pitch down will be the amount the aircraft naturally pitches down to maintain cruise relative wind less the small amount of back pressure we apply to fight the tuck. In steep banks the aircraft will tuck or drop the nose significantly to regain cruise relative wind rather quickly and we may wish to delay that to again miss the canyon wall or floor. The energy management turn is dynamic not static. It is not an acrobatic maneuver where absolute controls movements are made.

The chandelle is a coordination maneuver requiring prescribed pitch and bank attitudes that do not take advantage of hydraulic and/or thermal lift. The crop duster turn, wing over, gunship return to target turn, or what I call the energy management turn does not require full rudder application unless the angle of bank requires it. It is dynamic not static. In this turn we do not cross the controls. The surviving FBI agent who flew with the accident instructor either did not understand the maneuver or the CFI did not explain and/or demonstrate it properly. When practicing energy management turns with steep banks, students often think they are crossing the controls when in fact they are not applying sufficient rudder for the speed of nose movement and are thus slipping.

Let's look again at La Manga and Cumbres passes. *See Map 2.* Venturi's Law explains why the southwest wind blows so strongly up both these passes giving you tremendous hydraulic lift on slopes oriented across the prevailing wind. The problem is that the stronger the wind the steeper will be the declination line on the lee side of any slope oriented across the prevailing wind. *See Figure 22.*

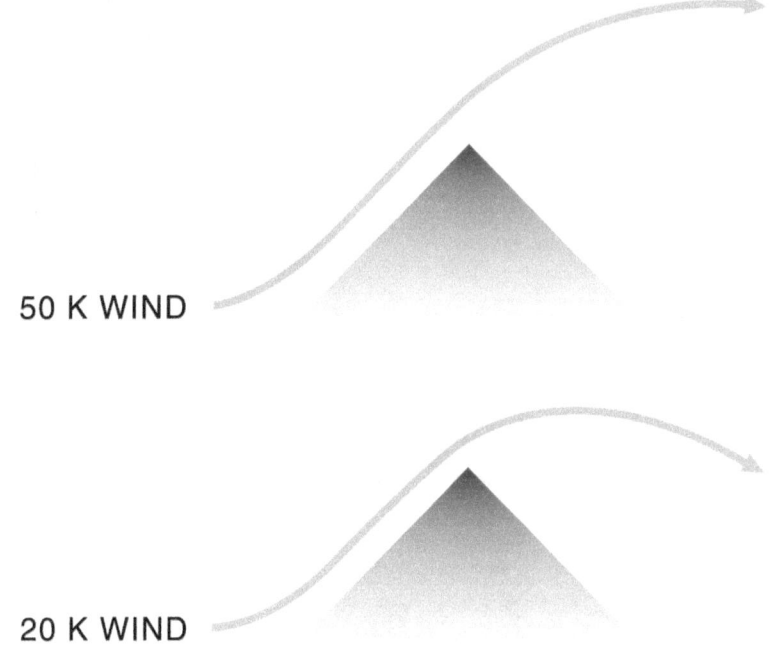

Figure 22

CONTACT FLYING

On La Manga Pass we may be able to ride the east ridge until southeast of Los Pinos, Colorado then jump across the valley to get lift on the south side of the peak under the "s" in Cumbres on the Denver Sectional. Once west of Cumbres we need to power descend to the valley floor at 8,000' or we'll have to buck much stronger headwinds at 10,000.'

Now consider our afternoon flight from Albuquerque to Gallup. Looking at Albuquerque Sectional we see V291 going directly west across Mount Taylor. We cannot go that directly except by thermaling up to get above the declination line. If the wind at West Mesa Airport is 30 gusting to 45, winds at 12,000' will be at least 60. This option might consume too much gas.

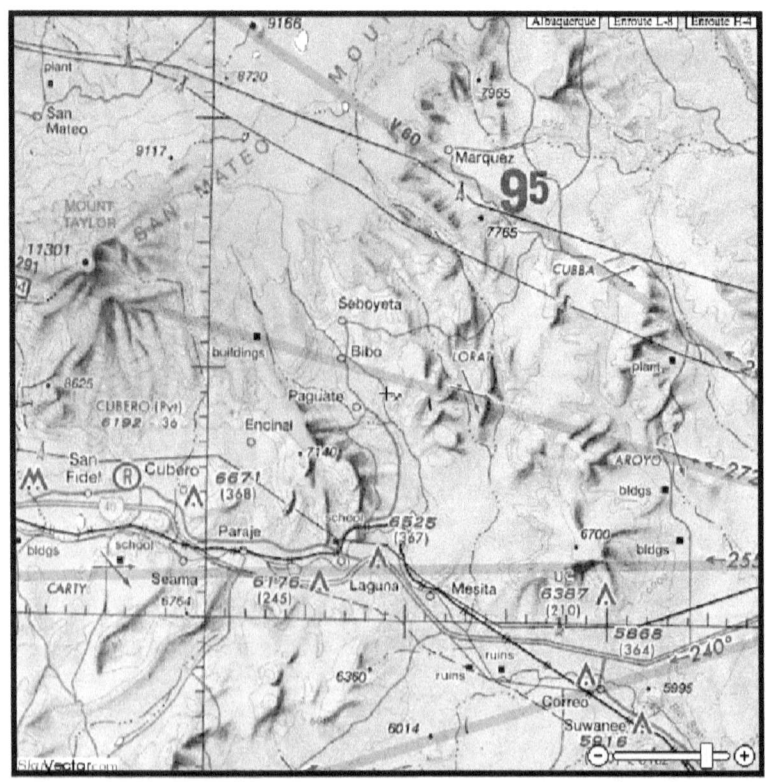

Map 3

So, we will improve our headwind situation and gain groundspeed by staying in ground effect as much as possible to the 6700' peak just north of V12, *(See Map 3)* the mesa west of the buildings by Arroyo intersection, or the ridge west of the plant on V60. Experience and instinct will help us find the best air. From one of these points we must decide whether to attack the valley along I40 or the San Mateo Mountains north of Mount Taylor.

The best route with southwest winds is simply to ridge ride up the north side of I-40 from Paraje to Gallup. With west to southwest winds, once past San Fidel you can turn up the intermittent stream over the high wire west of hill 8625 and drive straight up the southwest and west side of Mount Taylor to 11,000'. *See Map 4.*

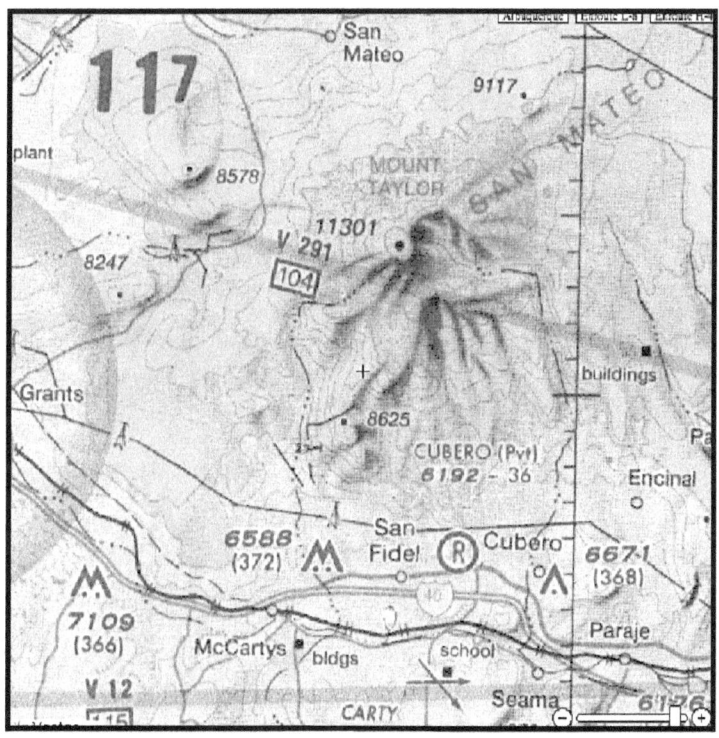

Map 4

CONTACT FLYING

If the wind is from the northwest, however, anything south of Mount Taylor will be dicey unless we luck into some wave action off the series of mesas running southwest to northeast from around Acoma Pueblo to the southeast slopes of Mount Taylor. I have generally found this unproductive and have spent excessive time ridge riding too far southwest to be economical. If we do go this way the ridge ride from Grants to Gallop staying south of I-40 is good. *See Map 5.*

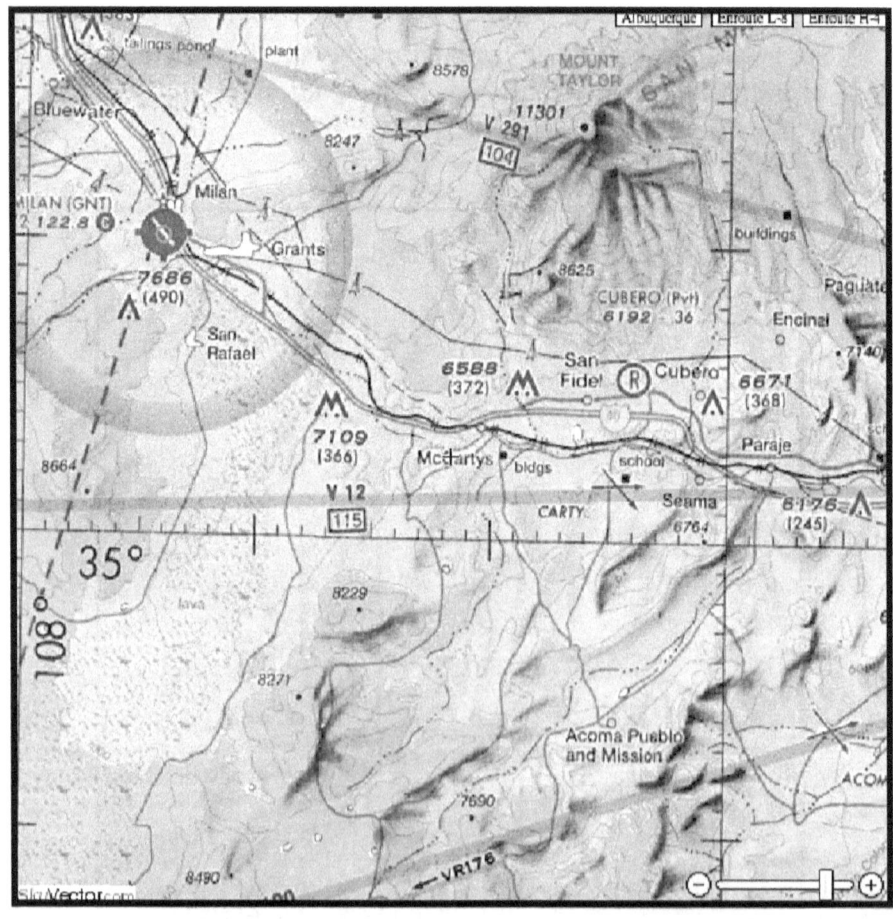

Map 5

With a northwest wind I usually find myself looking for wave action off one of the decision points mentioned earlier, *See Map 3* (the 6700' peak north of V12, the mesa west of the bldgs by Arroyo intersection, or the ridge west of the plant on V60) and trying to be hydraulically lifted to the San Mateo Mountains north of Mount Taylor. Kids don't try this at home. It looks easy and you will think your engine might carry the day, but the sheer walls and tight canyons of the mesa system from Paguate southeast of Mount Taylor north through Bibo, Seboyeta and Marquez will test your canyon turnaround skills. All these canyons are box canyons and oriented northwest to southeast. It's hard to find upslope air here but possible. Even when we do find upslope wind, the sheerness of the canyon walls often causes severe gusts and abnormal hydraulic conditions. The upwind side of rising terrain will always result in updrafts but if you get too near the sheer wall you may encounter compression shears so strong that they are not usable. For example we should expect updrafts over the wire from hill 7765 to Marquez. If we get too close to the mesa wall, however, we may have to hold the aircraft level, even while in strong updrafts, because the gust spread necessitates maintenance of zoom reserve in airspeed. Our reaction to entering compression turbulence, or lee side downdrafts, must be to turn sharply away from the wall and allow the nose to fall through.

It's hard enough to find good air on the lee side of a major system, so how are we going to deal with bad air on the upwind side? It requires patience, experience, and above all faith. The good air is there even if we have to follow the arroyo all the way around the San Mateo Mountains north of Marquez. Most days I have found good enough hydraulic lift to make the jump to the top of the mesa system. Once up on the flat San Mateo Mountain complex it is an easy mesa top run in ground effect (cooling the engine and avoiding the worst headwinds higher up) all the way to Gallup. *See Map 6.*

Whatever route we take, west of Mount Taylor, we must fly down low to conserve fuel. Staying high after high mountain pass crossings (on

CONTACT FLYING

westerly routes) strains the fuel reserves of slow aircraft. Water holes and fuel points are separated by vast expanses of desert in the West. Prevailing winds are westerly and strong every afternoon. We may forfeit half our afternoon groundspeed above 500' AGL. Gallop may easily be more than two hours away and that's not leaving any margin for having trouble making a pass. The higher we go the stronger the prevailing winds. We don't have to consult winds aloft charts to know this; it is obvious.

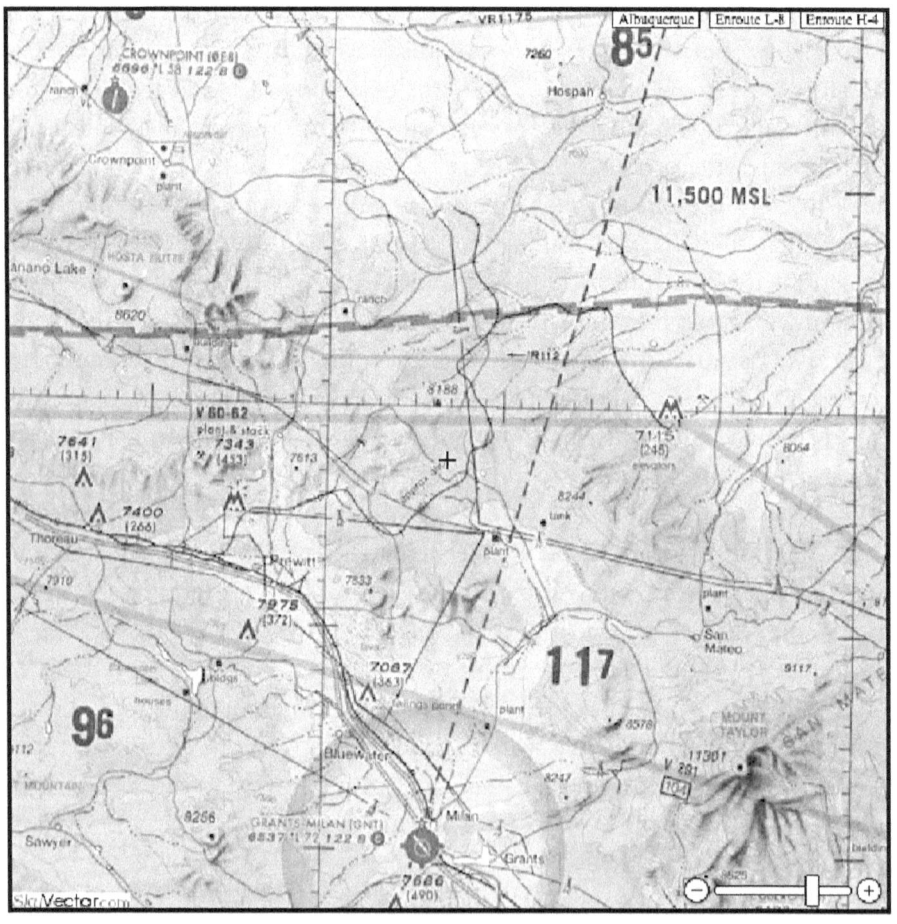

Map 6

Let's see how Sam is doing on his return trip to Gallup this afternoon. With northwest winds he elects to take V60-62 north of Mount Taylor. If he engine climbs to 12,500' thinking he needs 1000' above the highest terrain and applies the irrelevant hemispheric rule, he won't make Marquez in one hour. The remaining 65 miles in headwinds of 50 will take another hour. This assumes, however, there will be no reduction in airspeed to maintain altitude. There is a reduction in airspeed because holding a constant altitude will force him to fly slowly in downdrafts and fast in updrafts. Each time he encounters a 1,000 FPM downdraft he has to hold up elevator to maintain altitude (throttle is full to produce the 60% maximum power available at that altitude). This up elevator application slows him to 70 MPH airspeed. He is escaping the downdraft at a groundspeed of 20 MPH. Now let's say this downdraft lasts thirty seconds. If a Cessna 172 can climb 100 FPM at this altitude, and that is very questionable; he has lost 450' and can do nothing about it. But wait, you say, won't there be an updraft following? Yes and at 70 MPH and 1,100 FPM climb, Sam quickly returns to his 12,500' hemispheric altitude in less than thirty seconds. But what happens now? He pushes over to maintain cruise altitude and accelerates right through this gift from God. If he maintains 120 MPH airspeed he is escaping the updraft at a groundspeed of 70 MPH. Thus, Sam is flying through downdrafts at climb airspeed and updrafts cruise airspeed. This will cost him airspeed and altitude over time and distance.

How can we use thermal and hydraulic energy to build zoom reserve, which we may bank in altitude or airspeed? First, we ignore the hemispheric rule. It does not apply to small airplanes going west in the mountains because they cannot reasonably go above 3,000 AGL into strong prevailing westerly headwinds. Encountering the 1,000 FPM downdraft, let's push over to max structural (140 in a C-172). We will escape this downdraft at 90 MPH rather than Sam's 20 MPH. We will lose altitude faster than will Sam until compression ends the downdraft. Actually we did not lose altitude; we stored it in airspeed. On the other hand, we will fly slowly through the 1,000 FPM updraft. We will go through it at 20 MPH groundspeed instead of 70 MPH groundspeed. We will gain more than Sam but it will cost us airspeed. Actually we did not lose airspeed; we stored it in altitude. We will regain that airspeed, however, in the next downdraft. We will have a net gain in zoom reserve over time and distance. We can take this God given energy in the form of airspeed or altitude.

Map 7

I have experienced these phenomenal power surges in low powered aircraft many times. I once caught some ridge lift in a Taylorcraft DC-65 between Las Vegas, New Mexico and Santa Fe. *See Map 7.* For a time I was not sure I could even make the town of Bernal down Interstate 25 from Las Vegas. Winds were from the west and I wasn't finding any help for my 65 horse engine. After Bernal, however, I rode each north south ridge until high enough to jump to the next one. I ended up over Glorieta at 13,500' only because at Vne I could not descend in the tremendous hydraulic lift. I can't give you a number on that updraft as I had no VSI but it was good air and I thanked God for it. "If you have faith as a mustard seed, you will say to this mountain, 'move from here to there,' and it will move…" Matthew 18:20.

Chapter 10

THE CROSS COUNTRY

In 1996 I flew a Pawnee from Ontario, Oregon to Bainbridge, Georgia without electricity, GPS, or compass. I drew lines on sectionals and flew the resultant true angle to section lines for 3,000 miles. After a few hours cutting the same angle through successive sections the angle became imprinted on my brain.

Accurate navigation was possible, long before GPS, by reference to natural or manmade features at long range and by pilotage and reference to section lines. Not even a compass is required to navigate accurately anywhere in the US except in Spanish Land Grant areas (surveyed on cardinal headings of 45, 135, 225, and 315).

Accurate long range navigation is possible in the West using distant mountains and mesas. We must remember the shape of features on the distant horizon and remain low enough to see them. Anything higher than us will be profiled on the horizon. In the rest of the country towers and buildings can be used for long range orientation, if we stay low enough to profile them on the horizon. Remember that manmade features are not as reliable as natural features. New towers may confuse us and old ones may be taken down.

On any cross country we must pay attention to our true course drawn out in pencil on the sectional. The angle at which that pencil line crosses the longitude and latitude lines can be duplicated with the aircraft on the section roads and fence lines throughout the country except Spanish Land Grant areas. This angle is more accurate than a compass heading and needs no wind correction. If the angle of our flight path across these fence lines and section roads is the same as the angle of our pencil line on the sectional, we will go where the pencil line goes.

We must learn to sight down the true ground track of the aircraft. In a crosswind we will not arrive at the point between our legs unless we slip. While crabbing into the wind we need to develop a feel for where the aircraft is going rather than where it is pointed. If we estimate crab and pick points we will eventually get it through trail and error. Soon we learn to estimate wind angle and velocity by how it is affecting our flight path across section roads and fence lines. The wind correction angle is the angular difference between what we see between our toes (don't look over the prop in side by side aircraft) and where our butt actually will end up. Once we have sorted this out by trial and error we reverse think to the wind direction and wind speed that had to have caused this difference in where we were pointed and where we went. Now we project that information forward to estimate where to point our toes in order to arrive where we wish to go.

Setting three early checkpoints will allow us to establish the angle quickly. If we use pilotage to hit these first three we can establish the correct angle for a baseline. With further wind changes we will have to modify the crab angle but not the fall line (true course).

We ignore where the nose (between our legs) appears to be going. Instead we estimate the track of our butt across the earth. Once comfortable with where the aircraft (our butt) is going across the section lines, navigation by distant feature and/or section line will be no problem. It is more accurate than VOR, which is not usable at small aircraft altitudes in the mountains anyway. It is more accurate than dead reckoning. It has been replaced with GPS but only so long as electrical power is available.

When flying cross country into a headwind, we do best at 500' AGL or in ground effect in the open desert. This lessens the negative effect on groundspeed and gets us close enough to see road signs and water towers. Nothing gives us peace of mind like a water tower with the town's name painted on it.

CONTACT FLYING

Cross country techniques in the mountain and desert terrain require more flexibility, creativity, and courage (lack of fear of sanctions). On two trips from east to west I had to get creative with gas. Because of greater distances and strong winds every afternoon, we find ourselves sometimes laying up in unapproved landing zones in the interest of safety. As always we could have chosen not to have gone. By the same token we could choose not to fly.

While ferrying a Pawnee Ag plane back to Las Cruces, New Mexico from Missouri I found myself watching trucks walk right on by my aircraft in strong headwinds. By the time I reached Roswell, New Mexico I was in a part of the country where airports with services are at least one hundred miles apart. The Pawnee burned 15 gallons per hour and held 40 gallons usable. That was just a little over 2.6 hours with no reserve. With both spreader and spray boom attached the Pawnee trued out at 90 MPH. With a fifty MPH headwind I could make about 100 miles safely. I had four five gallon plastic gas cans in the hopper but I also had no radio. I landed on the road east of Cloudcroft and put the 20 gallons in. When over the road from Alamogordo to El Paso I decided it would be tight getting to Las Cruces so I decided to land in a dry lake by the road and hitch a ride into Alamogordo to get gas. As soon as I got down and out on the road a pickup stopped. I asked him for a ride into Alamogordo for gas. He said he had five gallons he would trade for my can and ten dollars. I bought his can of gas, gave him my can, went back to my plane in the desert, poured the five gallons into the tank, and took off for Las Cruces.

On another ferry from Missouri to Las Cruces I ran into a snowstorm east of Cloudcroft. I landed on the road, secured my Pawnee, and called my wife from the nearest ranch house. The only person at the ranch house was the maid. She did not speak English but provided me with a phone via sign language and my limited Spanish. Once on the phone, my wife assured her in Spanish that I was not one of los banditos Americanos. My wife hauled me to Las Cruces and I got a fellow instructor there to run me back to the Pawnee with his Cessna 150 the next day. We crossed Dona Ana Range below 500' AGL because the Army had all airspace above 500.' In the middle of Dona Ana Range a Crusader (Old Air National Guard Jet Fighter/Bomber) made a practice gun run on us. I had learned Army helicopter tactics to deal with such attacks. I turned into him to force him to increase his dive angle. This made him have to break off to prevent flying into the ground. After two more attacks where I turned into him (my buddy was freaking out) I called on 121.5 and asked, "Crusader over Dona Ana Range, what are you doing out here today?" He broke off and headed north toward Kirtland Air Force Base in Albuquerque, where Air Guard Crusaders were stationed. I explained to the instructor with me that the Guard pilot was just messing around.

On a trip back to Tohatchi north of Gallup, New Mexico after a drill weekend, I encountered strong headwinds in my Champ. I stopped at Grants to get fuel but they closed at five. After leaving a note and two dollars on each plane, I drained three gallons total from several quick drains in several aircraft and made it back to Gallup after dark. I had no electrical system in the Champ so I stayed low and circled looking carefully for fast aircraft before landing. In the days before radio click-on lighting you could get in a bind at night with a -8 engine (no battery, starter, or generator).

In the middle seventies I worked for spray operators in the San Louis Valley while my wife was working on her master's degree at Adams State College in Alamosa, Colorado. Continental Express had a daily flight into Alamosa that did not land when the temperature was over 90. The FBO caught me one day just before Margie and I were going to take off for Gallup in our Cessna 175. Bonnie said a timber salesman from Flagstaff, Arizona hadn't been picked up by Continental and needed a ride to Gallup where he could get on Frontier. I agreed to take Mr. Bingiman to Gallup. He was dressed in a suit and was carrying a briefcase. Margie got into the back seat and Mr. Bingiman sat up front with me.

CONTACT FLYING

Not far out of Colorado Mr. Bingiman asked if we would take him to Flagstaff for a reasonable fee. Margie agreed and I headed for Flagstaff. I had enough fuel to get there with a legal reserve but no extra. While checking weather from Winslow VOR I noticed that I was loosing volume. As the volume decreased, even though we were flying toward the station, I assumed I was loosing my receiver. Flagstaff was a tower airport so when I didn't receive a reply to my call, I transmitted "in the blind" that I could not receive but expected that I was transmitting and would need a green light to land when I arrived.

When we arrived at Flagstaff there was an overturned Cessna on the one and only runway. I requested that I might land long. Red light. I requested that I might land on the taxiway. Red light. I admitted marginal fuel but was unwilling to declare an emergency as I had been in hot water with Albuquerque GADO (General Aviation District Office) recently. Red light.

While all this was going on I had noticed Mr. Bingiman clutching his briefcase closer to his torso. I explained to him that we couldn't land at Flagstaff but we would try for Williams and land on the road if need be. We were on empty by the time we were over the Navajo Pipeline power station twenty miles east of Williams. I landed on the frontage road by the station.

Mr. Bingiman's wife came out to get us and after fueling the plane with gas from cans, we spent the night with the Bingiman's. Margie and I had just about figured out his concern with the briefcase as every coffee table and window sill was covered with old bottles. He told us at supper about his purchase of a very old and expensive pill bottle on this trip and yes he was worried about breaking it if we had crashed.

Chapter 11

AGRICULTURAL OPERATIONS

In 1995, between the first of June and the last of August, I sprayed 23,000 acres of corn in Furnas County Nebraska with a 235 horse powered Piper Pawnee agricultural aircraft. The numbers of people fed either the corn itself or the hamburger and steak produced from the beef that was fattened with 23,000 acres of corn is hard to fathom. This work was accomplished by an aircraft that was thirty one years old and by an older pilot who was paid seventy five cents per acre.

Agricultural operations include both the spreading of dry seeds, poisons, and fertilizers and the spraying of insecticides, herbicides, mitisides, defoliants and surfactants. It used to involve only spreading DDT, a dust formulation of the poison, and is still referred to as crop dusting. Like any aviation business the planes get bigger, fly higher, and fewer pilots are needed. New Air Tractors cost one and one half million dollars, have jet engines, and use drift retardants that help carry the spray down before it turns to dust but the objective to kill bugs, remains the same. The kids at the spray school often ask, "Why do those big yellow planes spray at 50 foot instead of 6 inches like you? Is it to get a better spray pattern?" I tell them, "Boys, the height a man sprays has nothing to do with spray pattern; it has everything to do with the hull value of the aircraft."

Prior to spraying a field we perform a high recon at 100' and a low recon at 5'. On the high recon we fly around the field to determine what obstacles are in the field and on its borders. We are especially concerned with wires. We want to know how every wire gets into the field and how it gets out of the field. We investigate every wiggle in the power line. We look for changes in direction because this requires a guy wire to keep the poles from leaning toward the inside of any turn. We want to make a

CONTACT FLYING

mental note of any wire than stops partway through our field whether in the field or on the border, especially if on the upwind side of the field. I once thumped the tire of a Pawnee on a phone cable that came halfway down the border of a field from the upwind side. Working crosswind and into the wind I did not arrive at this big black cable until halfway through the field after making many swaths where it was no factor. I had forgotten it on the important first swath where it definitely was a factor. Also there will be a guy wire at the end of a power line run, whether it ends at a house, pump, or goes underground.

On the low recon we get into the field at 5' AGL and look for anything that sticks up. Anything that sticks up into the horizon will be silhouetted on the horizon unless a higher object obstructs the horizon. In that case we must carefully pick out the obstacles in the field and on its borders against the further obstruction to the horizon. During the low recon we evaluate each wire as to whether we can fly under it or must go over. We also evaluate the slope of the field. We want to know if we have a steep slope to climb (this could slow us too much for the pull up with a full load) or if we have a downhill slope into tall trees which could cause us to pull up late with greater speed which could result in a mush or stall. When in irrigated fields we evaluate the height of dikes and look for stair stepped fields. I once hit a dike (called a border in New Mexico or West Texas) with the wheel of a Callair because I had not noticed that the adjoining vegetable plot was two feet higher than the one I was spraying in at 6" AGL.

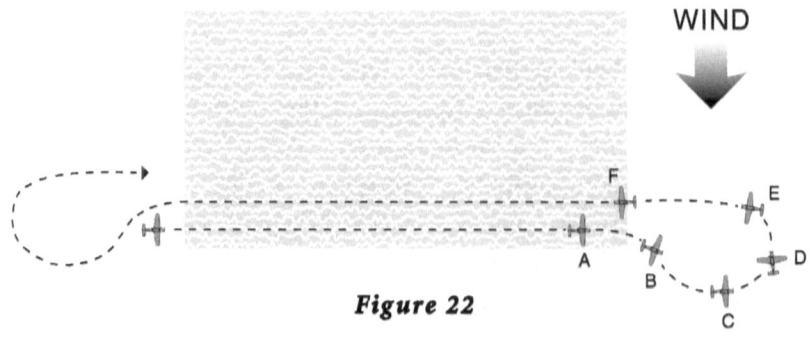

Figure 22

All Ag work is done crosswind and into the wind if possible. *See Figure 23.* In the field we keep our row alignment and spacing by observation of row, flagman, automatic flag, object at the end of the run, SATLOC (GPS with an expensive computer driven and user friendly interface), etc. When we get toward the end of the run we add full power, throw a flag, and shift our reference from the bottom to the top of any obstruction, pull up, and shut the spray valve off. The hand follows the eye and, without sacrificing energy by jerking the ship up, we zoom just over the obstruction. At the end of each spray run (A), just like at the end of a good takeoff run, the extra speed provided by ground effect helps us zoom up over the wires or trees. First we clear the wing (no fence, wire, tree, or other obstacle under it) and turn downwind (B). We deliberately choose to turn downwind first while we have zoom reserve in the form of airspeed. The angle here depends upon the strength of the crosswind but we want to get downwind quickly while close to the field. Whether we are still pitching up, level or even descending at point B depends on the amount of zoom reserve remaining and will differ with different aircraft, conditions and loads. In radial or jet powered aircraft we may pitch up to 500'. In Pawnees, Cessnas, or Callairs we will be lucky to get 150'. It all depends on zoom reserve, buoyancy; the sights, sounds, and kinetic feel of flight. Like the S turn the amount of bank depends on the wind. Unlike the S turn the amount of bank and pitch up depends on load, temperature, humidity, gust spread, etc. At point C, however, we turn back upwind allowing the nose to fall through. We need to complete the turn in sight of the field (E) and we need to level the wings prior to the pull up (F) to establish another three feet altitude in the field. This also helps to regain airspeed by point F where we also throw a flag.

The pull up and turn portion of the turnaround (B through D) will require much supervised practice. If we pull up too rapidly we load the aircraft and waste energy. If we pull up too slowly we may not be able to stay within visual range of the field. If we turn too rapidly we load the aircraft wasting energy. If we turn too slowly we will move too far from

CONTACT FLYING

the field and mush or even stall on top. If we do not complete the turn by E we will hook a boom or even a wing trying to get back to the proper row. If we do not complete the turn early we will load the aircraft late near the ground (still turning) and perhaps cause an aileron to go out (stall a wing). We must complete the turn prior to the pull up out of the dive so that we load the aircraft wings level. We need to complete the turn early to allow any necessary backpressure at E to prevent the nose from falling through too far. Shallow banks in this work will get you killed.

The crop duster turn, gun run, old lazy eight, or what I call the energy management turn is a ground reference maneuver. You've got to put the aircraft in the proper position over the ground to miss obstructions and keep the wings out of the ground. Like dancing we must assimilate the proper rhythm of the maneuver. Unlike dancing this rhythm will never be the same but will constantly change as load, wind, and conditions change. We are learning an art not a science.

As we come around to line up with the thirteenth row upwind from the last flag, we may need to hold a little back pressure to prevent a very steep dive (E). In every turnaround management of pitch down will depend on how much zoom reserve was used up in the pull up coming out of the field. The slower the ship at the wingover point the greater will be the pitch down as she tries quickly to regain cruise relative wind. If we allow too much pitch down we accelerate too rapidly in a very steep dive. If we hold too much backpressure she will mush down never regaining zoom reserve in the form of airspeed. Supervised practice is the only answer to learning this crop duster turn.

Once aligned with the proper row and anytime we are below 50' AGL we keep the wings level. If we need to adjust our heading, move over a row, or miss a tree in the field we push the nose around with rudder while holding opposite aileron sufficient to keep the wings level. If we allow the aircraft to bank we will drag the boom in the crop or even cartwheel. This same technique is used in a proper approach to landing. Once aligned with the runway centerline we set the wing level or at the appropriate bank for a sideslip into a crosswind. We set the wing then make all directional changes with rudder only.

While in the field we must constantly move our head left to right as if on a swivel. This technique is also taught Army pilots in "nape of the earth" (NOE) flight. We tend not to see things that are straight ahead if we never turn our head to pick them up in our peripheral vision. To often sprayers who forget this technique hit a tree in the middle of the field.

Leaving or reentering the field we throw a flag on the non-crop area prior to the field or in the field depending on where it will show up best. Modern agricultural aircraft do not use flags as they have computerized GPS units called Satloc.

We handle obstructions in the field by throttling back a bit, pulling up over the obstruction, adding full power, and reentering the field on the other side. As you push over into the field beyond the obstruction, the engine will momentarily quit. Negative gravity has closed the carburetor float valve. Upon pull up after the jump the engine will restart as positive gravity returns the float to the normal position. This will keep us awake on those long runs in big fields.

When wires have no obstructions under them and are sufficiently high we go under them. This is easier and safer than going over but requires relaxation. If we concentrate on or even look up at the wire we may fly into it. The hand tends to follow the eye. We must concentrate on the row ahead and watch for traffic on the road. We know a wire is there because we see the poles.

Not all spraying and spreading is done back and forth in the return to target mode. We racetrack big fields and arrange the little fields in circuits where possible. Much of the time, however, the return to target is the only practical pattern.

CONTACT FLYING

Spraying high altitude fields require a much flatter turnaround. Turns are much wider than at lower altitudes but there are fewer obstructions in the desert. The pull up to leave the field may sap all zoom reserve so that the turnaround involves a long flat turn just keeping the down wingtip going between obstructions and out of the ground. It is similar to having to make a downwind return to target.

While teaching History and English on the Navajo Reservation I sprayed alfalfa weevils during summer vacations around Durango La Plata Airport between the Florida and Pinos rivers. I limited my loads to 100 gallons and worked mostly off the airport or section roads. Later I sprayed vegetables, cotton, chili, alfalfa, and pecans in the Mesilla Valley south of Las Cruces, New Mexico and the South Valley from El Paso to Fabens, Texas from March to September.

In 1996, while teaching crop dusting at Ag Flight in Bainbridge Georgia, I ruptured the disk between station L4 and L5 in my backbone. After surgery I returned to Georgia to continue teaching but had further back problems and had to quit. After one year down I began flying pipeline, which was easier on my back. I still teach crop dusting and help former students out when they have a bug run, but my body will no longer hold up to very many long days. Pipeline work requires long hours in the saddle but will not kill your back like pulling back hard on the stick a thousand times a day.

Chapter 12

PIPELINE PATROL

For several years I have patrolled Plains All American Pipeline's Basin 24" crude oil pipeline (formerly a Shell pipeline) and their 10" crude oil pipeline (formerly an Enron Oil Transportation and Trading pipeline then a Link Energy pipeline) from Colorado City, Texas to Wichita Falls Texas. During most of this segment on these pipelines, they ran side by side in the same right of way. There were several short segments on this run however where the two pipelines separated as much as one quarter of a mile. In one of these separations, I found oil out on the ground on the EOTT 10" crude oil line.

In today's high technology, glass cockpit world it is important to understand what I was doing and more importantly what I was not doing in order to find this oil out on the ground. I was looking out the left front portion of the front windscreen on a Cessna 172. This task would have been easier in an older aircraft because there would have been more windscreen and less instrument panel and glare shield. Higher instrument panels decrease our ability to patrol pipelines and also to see and avoid other aircraft.

What I was not doing was of little relevance in the old days but has become increasingly significant today. I was not observing the directional gyro nor adjusting it so that it would agree with the magnetic compass. I was not observing nor was I adjusting the artificial horizon. I was not observing the rate of turn indicator. I was not observing the trim ball because I adjust trim using the rate of nose movement verses the angle of bank as observed outside the aircraft. I was not observing the altimeter as my naked eye is much more accurate at two hundred feet. I was not changing radios or GPS settings; neither of which have any value in actual patrolling. I was not observing the airspeed indicator. I use groundspeed

CONTACT FLYING

adjusted by estimated wind speed observed outside the aircraft and the kinetic feel of things, stick position (feel not sight), relative wind noise, lack of gas smell, and the relative wind pressure against the control surfaces transmitted through the wheel. I was not observing the GPS user waypoint track which I only use to learn a line and never to patrol it.

I was looking outside the aircraft and I found the oil. When I look outside the aircraft I find equipment, construction projects, pipeline exposures, dead animals, erosion over the pipe, survey crews, downed trees, logging operations, drill rigs, dead vegetation, and unusual activity on or near the pipeline right of way. I also see trees, crop dusters, power lines, towers, ultralights, and unidentified objects either rising up from the ground or flying just above it. When I look inside the aircraft I find none of these things.

Visual reconnaissance, whether in the jungle looking for Charlie or on the pipeline, requires constant vigilance of the target area. In our search for leaks, and the equipment that may cause them, we must continually position the aircraft so that all of the right of way is in view. This requires both lateral and vertical aircraft position adjustments depending on the situation and the lay of the land.

There are many aircraft types and techniques in the pipeline patrol business. Most patrol aircraft are slow and high wing. Most pilots fly well right of the pipeline and high (500'-1000') looking out the left window under the wing. High wing aircraft and observation only out the left window may be required for those with 20-20 or worse vision. For those with better than 20-20 or who are willing to fly very low, neither a high wing nor observation out the left window is necessary. The easiest and most effective way to patrol a pipeline is to fly low enough to see the mile markers out the left front or right front window.

When the right of way (ROW) cuts a swath through dense forest, two aircraft positions allow us to see the ground at the base of the tree line.

We could be up at 400' looking out the left side in order to view the entire ROW. On the other hand we could be low and just over trees on the right edge of the ROW. When over farmland and prairie we need to be much lower to get a very slight angle on the pipeline marker poles and mile markers. Any altitude from pole height to 200' works here. We stay slightly to the right of the ROW because we sit in the left seat (unless we have single or tandem seating). We stay in close enough to keep the necessary heading while looking well down the ROW.

If we can see 20' or further where the average person requires a distance of 20' or less, we will be able to read the mile marker at 100'. If this is not possible we need to lay off further right and bank toward the mile marker just as we go over. This look straight down technique is also required to judge between black water and oil. When we look straight down through the liquid we may see the bottom or at least see deep into it. If it is crude oil we will not see the bottom or even into the pool. That stuff is really black. Fuel oil or gas has a rainbow appearance on water. Natural gas leaks will bubble in water. A caution on crude oil when the wind is blowing; it drifts downwind on water. The upwind portion of any pond, lake, or river will be clear of all crude oil so you've got to look around.

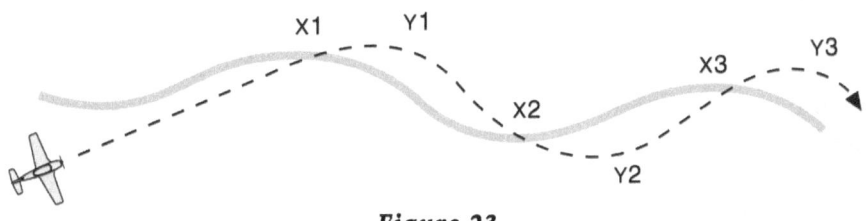

Figure 23

We see only what we are looking for with intent to see. We seldom see what we don't intend to see. We must intend to see. We must lean into our work. If our back does not hurt at the end of eight hours leaning into our work, we have not been patrolling. Rather we have been flying a route.

CONTACT FLYING

If the line is very crooked we need to stay very nearly over the line so that we can respond to the rapid turns. We look down out the left front window when the line is straight or turning left. But when the line turns right we need to cross over the ROW at x1 well to the left side so that we may look out the right front window in a much steeper right bank at y1. *See Figure 23.* Before reaching the crossing points (Xs) we must mentally mark where we last saw the ROW out the left front window and then pick up that point out the right front window when we put the right wing down. We continue looking and turning right to x2 and again marking something on the ROW so that we can pick it up out the left front window when we straighten out or put the left wing down. The turn at y2 need not be as steep as the turns at y1 and y3 because we are looking out the left front where we have less engine cowling to look over. As always wind direction must be considered in all turns.

This technique will work on snaky crude oil gathering lines. Gathering systems bring oil from the producer's pumps to the gathering line and then to large collection tanks. The turns will be very tight amounting to steep Dutch Rolls. With each roll you look down out the appropriate side of the front window.

Many pipeline patrol pilots fly only on the right side of the ROW looking down out the left door window. This rigidity forces them to stay well right of the ROW on crooked lines so as not to inadvertently fly over the ROW and loose sight of an area they had not yet seen. When this far away from the ROW they have to be very high to see down the line and not get lost. Often they cannot see where the line goes next and have to memorize structures and terrain features. Reporting a spot requires flying back to the point to find a mile marker as the number cannot be seen from far away.

Dead areas are missed on crooked lines and patrol is impossible on a line with many square turns followed by straight stretches *See Figure 24.* In the right turns at x1 and x2 the pilot on route A will miss the entire corner as it will be blocked by the left fuselage and belly of the aircraft. Proper coverage of this dead area would require 270 degree turns at x1 and x2 while looking out the left side of the aircraft. On the other hand flying this line as on route B crossing the line before the steep right turn would be less hassle and provide better coverage. When we turn right while sitting in the left seat, we must position our aircraft so that a steep right turn is necessary to again cross the line. Only in a steep turn can we see out the right front window across the panel.

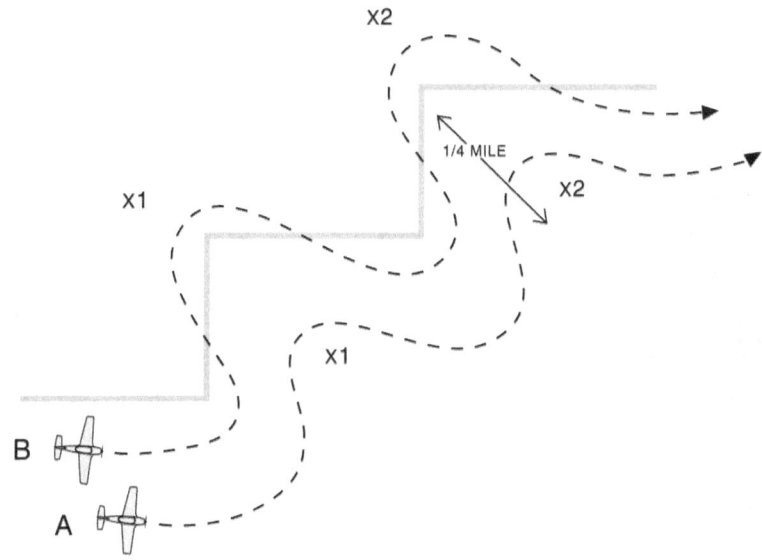

Figure 24

Instrument panels are unnecessary and dangerous (because they limit outside vision) in patrol aircraft. This unnecessary and dangerous condition increases with the manufacture date of the aircraft. As manufacturers sought to increase sales with increased utility through increased space for instruments and other gadgets, their design engineers

CONTACT FLYING

increasingly pushed the panel upward decreasing the pilots forward and downward visibility. While this tradeoff increased available space for more radios and instruments it increased the risk of midair collisions and decreased the utility of these aircraft for reconnaissance work.

In a strong left crosswind we will find ourselves having to look over the right side of the nose when patrolling close to the line. A simple solution is to land and change to the right seat. Now we can patrol out the right side of the front window just as effectively as we normally patrol out the left side of the front window. The strong left crosswind is really a problem for those who insist on always flying right of the ROW. Good coverage is very difficult because one has to constantly shift ones vision from over the right nose (to see where the pipeline and aircraft are going) to well behind the aircraft (looking out the left window) where the actual patrol is conducted. This is not only difficult but also dangerous.

All sightings of equipment, construction, dead animals or vegetation, erosion, log jams on exposed pipe in streams, compromised risers or valves or pig traps, fires, survey crews, and leaks must be reported to the oil company by mile marker. Lines are marked in one mile, two mile, three mile, or five mile increments. Mile markers are from six inches square to 12"x 24" rectangles. It takes some mental gymnastics and skillful ground reference flying to mentally record the sighting (type and number of equipment, distance from ROW, active or not active, etc.), visually acquire and descend on the next mile marker, mentally record the number, calculate the distance back to the sighting, confirm the mile marker by visually acquiring and descending on the following mile marker, and finally recording the sighting on a line sheet. When we start writing before we have done all this we will not be able to confirm the mile marker number and stay on the line. It is much easier and more efficient to do this at 100' than at 1,000'. Flying high requires the patrol pilot to turn around and descend on every sighting to confirm the mile marker location of that spot.

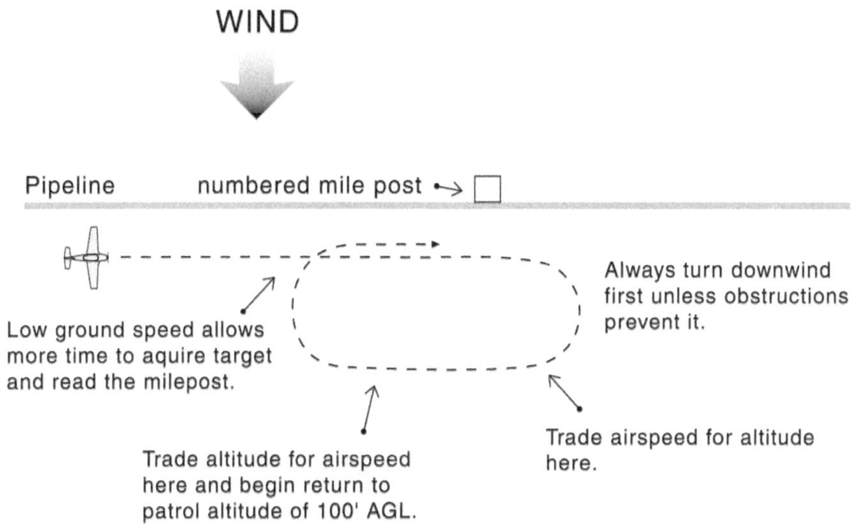

WIND

Pipeline   numbered mile post

Low ground speed allows more time to aquire target and read the milepost.

Always turn downwind first unless obstructions prevent it.

Trade altitude for airspeed here and begin return to patrol altitude of 100' AGL.

Trade airspeed for altitude here.

Figure 25

The turnaround, to check a spot or find a mile marker, is the same as the crop duster turnaround or gunship return to target *See Figure 25*. We turn off line downwind, pitch up to slow down for the turn to return to target, allow the nose to fall through in the turn to regain speed and lose the gained altitude, hold back pressure as necessary to prevent an extremely nose low attitude, level the wing, and walk the nose onto the target with rudders.

Getting back on line after fueling requires a similar technique. As we approach the line upwind with slow groundspeed we will be able to turn directly onto the line. We just pitch up to slow the approach speed then allow the nose to fall through onto the ROW. Approaching the line downwind will usually require a 270 degree return to target. *See Figure 26*. We pitch up when over the ROW then turn away from the desired direction of travel. When sufficient zoom reserve has been transferred from speed to altitude, we allow the nose to fall onto the ROW now exchanging altitude for speed going down the ROW. This is just like getting on the interstate highway at a cloverleaf intersection.

CONTACT FLYING

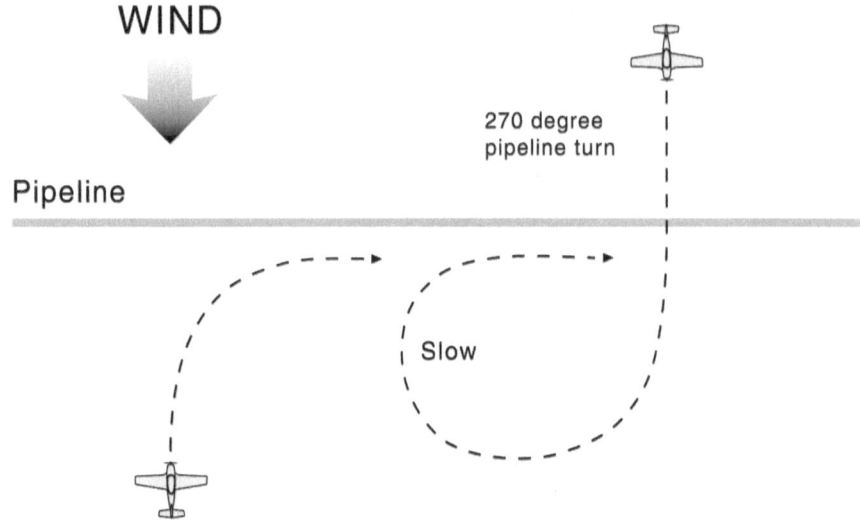

Figure 26

Department of Transportation (DOT) requires any interstate oil pipeline to be shut down on the 21st day it has not been flown and to remain shut down until flown. This means that most lines in the Midwest are flown every week. Patrol contracts run about one dollar per mile on average. With such low profit margin my boss cannot afford to keep me if I average much less than 95% completion on a weekly basis. Were I to quote regulations and not fly in marginal weather I would not keep my job. Bad weather, then, is the most serious threat to the pipeline pilot.

A technique useful to flight check marginal weather safely is the crop duster turnaround (without the pitch up in very low ceilings). If we stay in ground effect until zoom reserve in the form of airspeed is established, we may use this energy to make a level 270 degree return to land in the opposite direction of our takeoff. *See Figure 27*. We use the energy management turn as much as possible but avoid going into a potential 100' ceiling. Thus we often have to make a more dangerous level turn.

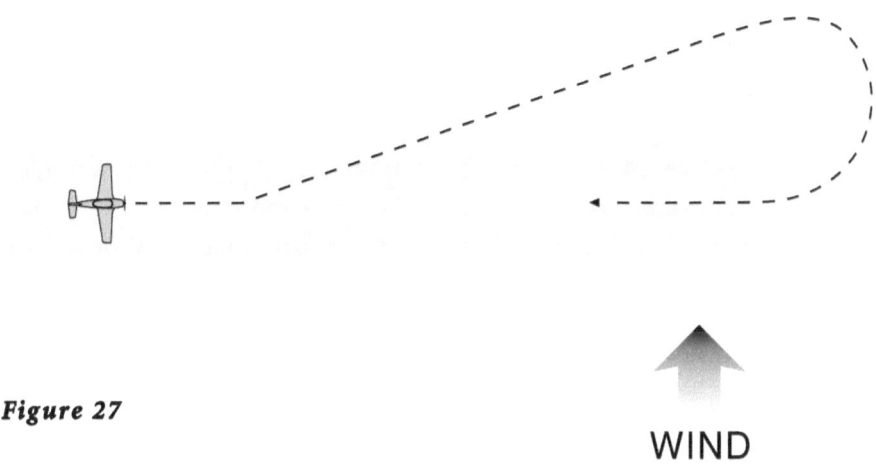

Figure 27

WIND

Wind must always be taken into consideration. Where possible we turn downwind first to allow an upwind (low groundspeed) turn of base to final. This way we can safely evaluate the flight weather without committing to fly the entire pattern in it.

As we fly down the line toward deteriorating weather we consider airports near the line as rally points. Should things continue going downhill we make a turnaround and retrace our route on the line while setting up GPS, sectional, or user waypoint precautionary landing fields. Remember that electronics usually fail when most relied upon and reading a sectional while flying low level in bad weather is dangerous. The nearer we land to our last work the sooner we will be able to observe any weather changes out there. If we need a new big picture, however, we need to return to a field with computerized weather.

Flying pipelines in Class C and B airspace has given me insights into how small aircraft operating VFR can cooperate with ATC to maximize comfort levels for pilots and controllers. I flew lines that crossed major airports at Jackson, Memphis, Indianapolis, Cincinnati, Chicago Midway, Navy Ft. Worth, and Ft.Worth Meacham.

CONTACT FLYING

Because of an anomaly in the regulations, B airspace is generally more user friendly to small VFR aircraft than is C airspace. C requires radio contact before entry into the outer ring while B does not exist for pipeline patrols until five miles from the primary airport at our altitudes. Approach controllers certainly appreciate our not calling them, nor should we, when well below their airspace. Approaching Ft. Worth from the west I called Navy Ft. Worth five miles out, crossed the south end of their long north south runway, was handed off to Meacham and crossed their north south runway midfield. After the tank farm east of Meacham I circumnavigated DFW five miles or more out at 500' AGL. At Cincinnati Blue Ridge I contacted tower five north or south, crossed the east west runway, and then crossed the western north south runway taking a jet fuel line to the tank farm between the parallel runways. Indianapolis was the nicest C airspace I flew in. They overlooked the approach call requirement and allowed me to call tower five miles north or south. I crossed the north south parallel runways just south of the tower. All other Class Cs (Jackson, Chicago Midway, Longview Texas, Shreveport, Albuquerque, Tulsa, and Springfield, Missouri) and Memphis Class B approach controllers required me to climb up high enough to contact them further out. This procedure required me fly the line low level until twenty miles out. Next I would have to reverse course and climb, while flying back down the line, to initiate radio contact with Approach Control. Once radio contact was made I could legally reverse course and descend to continue flying my pipeline. Until controllers became aware of my needs (I flew these lines weekly) they would fuss about loosing radar contact. The accepted procedure is to request negative radar. It is much nicer in Class B and user friendly Class C to just keep on trucking until you hit Class D. Many times the tower controller at Cincinnati, where I was not required to contact Approach Control, replied to complaining airliners with their proximity devices squawking, "He's fifty feet off the trees. He's not traffic."

The point I'm trying to make is that busy airspace, whether B or C, is high airspace. For small and slow airplanes, staying out of the mix can be safer than demanding your pound of flesh. The easiest way to enter an unfamiliar busy airport is IFR. If you're flying a Champ with a 65 horse -8 (no starter or alternator) or a beat up 172, that may not be an option. In that case staying low and out of the mix is much easier on the nerves if not safer.

When patrolling in the mountains, mountain techniques have to take precedence over pipeline techniques. Pipelines do try to find the least imposing terrain and therefore we can usually utilize both techniques. We do, however, find ourselves downwind of significant mountains in high winds on windy afternoons (think every afternoon). This can be dicey when flying west into higher terrain. I flew a right of way east of the Sandia Mountains that had this problem. I maintained normal pipeline altitude (200' AGL) from Moriarty Airport to Sandia East Airport to Sandia Park and then used the ridge east of Placitas to get to Placitas. From Placitas it was downhill to the Rio Grande. I crossed the Big River (Rio Grande) near Santa Ana and continued northwest to the mine south of San Ysidro and then on up to Durango crossing the Continental Divide on a high flat desert.

Because the right of way is generally remote from built up areas and very visible from any angle, we may take any thermal and ridge lift available to climb to as high as 1,000 AGL and still see equipment or problems. We will have to descend to read the nearest milepost in order to call the problem. We then use mountain techniques to get back up over the next pass. This would seem a real burden but is not because of the isolation and lack of reports necessary. While I call an average of 25 spot reports on midwest pipelines I seldom call more than two or three in the same distance on western lines.

Like any cross country in the West, patrolling western pipelines requires laying up from time to time to save time and fuel. Warm fronts are uncommon in the West because they are rapidly occluded by fast moving cold fronts. We can usually land on an isolated road or dry lake bed and wait weather out. I always carried a sleeping bag and tent but seldom needed to use them. The big problem for an old pilot with a bad back is getting the plane off the road and back on again to take off. You could wait until help came along, but I once slept on the wing of a Pawnee on a section road in western Kansas for two hours before a grain truck and combine came along. The farmer checked to see if I was alright, but assumed I was just waiting for the upslope fog to burn off.

Chapter 13

THE NEW FRONTIER

In 1994 I returned to Crane Missouri and the Ozark hill country of my early years. Having started flying at a very young age, I was known in these parts as a pilot. The best known pilot in Crane today is the owner of Hard Knock Body Shop. Perry (Perk) Vaught is without fear on motorcycle or in ultralight aircraft. While crashes were not uncommon in his motorcycle career, his ultralight career began with a crash. After watching his cousin make a few takeoffs and landings with his rebuilt Birdman ultralight, Perk parroted the very high angle of attack takeoff he had observed, stalled, and went straight in. Luckily the Birdman will only go 40 mph straight down at full power.

Perk came to me with a strange request. He wished to watch me fly his ultralight to determine how to fly it. Having never flown one and being afraid of heights, I was somewhat reluctant. Few can stand up to Perk's optimism, however. I made ground effect takeoffs and climbed to only fifty feet above the trees. After the second circuit Perk came running out shouting that he now saw what his problem had been and was prepared to fly. He made several fine takeoff and landings and is one of the most natural pilots I have ever met.

Technology and boldness has seemingly opened a new frontier in aviation. That new frontier is the airspace between the ground and 500' AGL. Ultralight aircraft have again popularized this wonderful space that spray and pipeline aircraft have been operating in all along. This airspace is the best airspace for learning to fly quickly because it gives the pilot a clear picture of the effect of his control inputs.

Technology also opened the old new frontier of flight within the clouds. William Langewieche is correct in his book, Inside the Sky, that the seat of

CONTACT FLYING

the pants type of flying I have been writing about is useless in the instrument environment. This is an area where the transfer of contact skills to instrument flying is not only useless but also harmful. Yet we do not enhance the instrument environment by denying the ground reference environment.

I live in another world from most of general aviation. Because I am below the horizon in their point of view, most pilots never see me. I am not traffic for them nor are they traffic for me. The only aircraft that share my sky are ultralights, sprayers, and other pipeline patrol aircraft. We are all in a see and avoid mode because we have nothing on the dash to distract us.

The area of potential conflict between the two frontiers is around airports. All aircraft, those flying high under instrument flight rules as well as those flying low level in contact with the ground, must alight from the ground and return to it. We who fly low level must avoid those who fly normal traffic and instrument approaches. We are at a great advantage in that we are low and we can see them easily above the horizon. We are at a disadvantage in that they cannot see us as easily. It therefore behooves us to see and give way to everything flying around. This includes turning to clear our tail before lining up with the runway on final when landing.

While this new frontier has the attention grabbing appeal of an extreme sport, it is as safe as high altitude flying. After all we have been operating in this environment for over a century now. Flying has been made safer by better certification procedures, better engines, and much experience with the same basic aircraft. Higher is not necessarily safer except that much weather is avoided with rocket powered aircraft. Simple airplanes are safe because they are simple. Wolfgang Langewiesche can teach you how and why the airplane acts in a reliable way in <u>Stick and Rudder</u>. Lycoming or Continental can put a reliable engine on it for you. Contact Flying and Aircraft can teach you how to find natural energy in the low level environment that will make up for low power. We can teach you to use natural energy and your wing.

Try staying below 500' AGL for an entire month, weather and mission permitting. Turn all radios off and cover all gauges except oil pressure and temperature. If your engine is small enough, tie the tail down and prop the engine for each start. Did you know you could actually smell and hear the fuel air mixture so critical to starting small engines? Get an old guy to teach you all the elements of propping. When I have trouble starting an airplane I always get out to use my nose and ears. I can't smell or hear an engine properly from the cockpit.

The high altitude world is performance and prestige oriented. Cessna 150 pilots dream of 172s, 172 pilots dream of Centurions, Centurion pilots dream of 402s, and 402 pilots dream of Citations. In Jonathan Livingston Seagull, Richard Bach says perfect flight is "being there." If this is true why not just get an inexpensive (when compared with flying yourself) airline ticket? The reason is that SUVs (driven mostly in cities) and airplanes (flown about 15 hours per year) are cool. Furthermore bigger is cooler. If we buy into this world we don't want safety and reliability; we want power!

When our objective is either recreation or actually making money with our aircraft, smaller is often better. Small aircraft experience tremendous diminishing returns when we increase engine power to achieve increased speed and altitude. Both speed and altitude can be our enemies when our mission is to observe the ground below or do a good job killing bugs. Look at the development of remotely controlled aircraft in military scouting and target engagement.

Part of the attraction of flying is to be in command of your own destiny. We can use old, simple aircraft to gain some independence. We can fly an older aircraft that has no electrical system and has never had one. We can fly low level to the middle of some metropolitan area. We cannot call center, approach control, or tower; nor do we have to. We can land at an uncontrolled airport. We can walk to the nearest restaurant and eat like a king.

BIBLIOGRAPHY

Hayes, Michael D., Certified Flight Instructor Oral Exam Guide, Aviation Suppliers & Academic Inc., 7005 132nd Place SE, Newcastle, Washington, 98059-3153, ISBN 1-56027-457-3, ASA-OEG-Cf 14

Welch, John F., Flight Instructors Pocket Companion, New York: McGraw hill, c 1997, 629.132W, ISBN 0070691924 (pbk)

Langewiesche, Wolfgang, 1907-   , Stick and Rudder: an explanation of the art of flying., New York: McGraw Hill, 1972, 629.1325L, ISBN 0070362408

Apocalypse Now (video recording) / an Omni Zoetrop production, directed and produced by Francis Coppola, Hollywood, Calif., Paramount Pictures, 1998, (c 1992)

The Postman (DVD video recording), Warner Bros. presents a Kevin Kosner film; a TIG production, Burbank, CA, Warner Home Video, c 1998

Butch Cassidy and the Sundance Kid (video recording) / Twentieth Century Fox, Beverly Hills, Calif., 20th Century Fox Home Entertainment, . c 2000

No part of this book shall be reproduced, stored, in any retrieval system, or transmitted by any means, electronic, mechanical, xerographic, audio/visual record, or otherwise, without the written permission from the author. While every precaution has been taken in the preparation of this book, Jimmy A. Dulin assumes no responsibility for errors or omissions. Neither is liability assumed for damages resulting from the use of the information contained herein. None of the material in this supersedes any documents, procedures, or regulations issued by the Federal Aviation Administration.

CONTACT FLYING

www.ingramcontent.com/pod-product-compliance
Lightning Source LLC
Chambersburg PA
CBHW020947230426
43666CB00005B/213